Assessing the Value of U.S. Army International Activities

UA
25
.A7959
2006

Jefferson P. Marquis, Richard E. Darilek, Jasen J. Castillo,
Cathryn Quantic Thurston, Anny Wong, Cynthia Huger,
Andrea Mejia, Jennifer D.P. Moroney, Brian Nichiporuk,
Brett Steele

Prepared for the United States Army

RAND ARROYO CENTER

The research described in this report was sponsored by the United States
Army under Contract No. DASW01-01-C-0003.

ISBN-10: 0-8330-3803-6
ISBN-13: 978-0-8330-3803-6

The RAND Corporation is a nonprofit research organization providing
objective analysis and effective solutions that address the challenges
facing the public and private sectors around the world. RAND's
publications do not necessarily reflect the opinions of its research clients
and sponsors.

RAND® is a registered trademark.

Photo Courtesy of U.S. Army

Published 2006 by the RAND Corporation
1776 Main Street, P.O. Box 2138, Santa Monica, CA 90407-2138
1200 South Hayes Street, Arlington, VA 22202-5050
4570 Fifth Avenue, Suite 600, Pittsburgh, PA 15213
RAND URL: http://www.rand.org/
To order RAND documents or to obtain additional information, contact
Distribution Services: Telephone: (310) 451-7002;
Fax: (310) 451-6915; Email: order@rand.org

Preface

This report presents a conceptual framework for assessing the value of U.S. Army International Activities (AIA) and describes the new Army International Activities Knowledge Sharing System (AIAKSS). Although a number of important steps have been taken in recent years to improve the planning and management of the Army's non-combat interactions with foreign militaries, the need for a high-level assessment mechanism is widely recognized to allocate AIA resources more efficiently, to execute AIA programs more effectively, and to highlight the contributions of AIA to the Defense Strategy, the Department of Defense (DoD) Security Cooperation Guidance, and The Army Plan. Based on extensive discussions with security cooperation officials in the Army and other DoD organizations, this report presents a global, program-oriented assessment approach designed to fit within a future DoD-wide evaluation framework that includes the assessment mechanisms being developed by the U.S. regional Combatant Commands.

This report is the final product of a two-year study entitled "Assessing the Value of U.S. Army International Activities." The study was sponsored by the Deputy Chief of Staff, G-3, U.S. Army (Headquarters, Department of the Army) and was conducted in RAND Arroyo Center's Strategy, Doctrine, and Resources Program. RAND Arroyo Center, part of the RAND Corporation, is a federally funded research and development center sponsored by the United States Army.

This document should be of interest to individuals in the government, academia, and nongovernmental agencies concerned with

the planning, management, and evaluation of U.S. security cooperation activities and programs.

The Project Unique Identification Code (PUIC) for the project
that produced this document is DAPRRX017.

For more information on RAND Arroyo Center, contact the Director of Operations (telephone 310-393-0411, extension 6419; FAX
310-451-6952; email Marcy_Agmon@rand.org), or visit Arroyo's
web site at http://www.rand.org/ard/.

Contents

Figures

Tables

Summary

This study reports on the development of a conceptual approach to assessing the value of the U.S. Army's non-combat interactions with other militaries. The first task was to take a fresh look at the objectives or "ends" of Army International Activities (AIA). Because the Army conducts security cooperation activities based on policies determined by the Office of the Secretary of Defense (OSD), the Department of State, and other governmental agencies, we reviewed all of the relevant national security documents: the National Military Strategy, the Defense Planning Guidance, the Quadrennial Defense Review, and OSD's April 2003 Security Cooperation Guidance (SCG), as well as the AIA Plan and The Army Plan (TAP). The result was eight AIA *ends* (listed in italics, below), which we embedded within OSD's three overarching security cooperation objectives (listed in bold, below):

- **Build defense relationships that promote specific U.S. interests**
  — *Assure allies*
  — *Promote stability and democracy*
  — *Establish relations*
  — *Improve non-military cooperation*
- **Develop allied and friendly military capabilities for self-defense and coalition operations**
  — *Promote transformation*
  — *Improve interoperability*
  — *Improve defense capabilities*

- **Provide U.S. forces with peacetime and contingency access and en route infrastructure**
  — *Ensure access*

The second task was to consolidate the large number of Army International Activities into a manageable set of AIA "ways" (see Table S.1). Because of Headquarters, Department of the Army's (HQDA's) interest in evaluating the progress of AIA as a whole, some degree of consolidation seemed appropriate. Furthermore, we were hopeful that if G-3 accepted our categories, we might gain support for their use within the context of the Army's Planning, Programming, Budgeting, and Execution System (PPBES). Currently, AIA are spread across several Program Evaluation Groups (PEGs) and a multitude of Management Decision Packages (MDEPs), which inhibits HQDA G-3 resource managers from making the best case for Army security cooperation activities.

Although the above list was developed before the SCG was published, the differences between OSD's categories and ours are minor. For example, OSD's category labeled "other" is largely identical to what we call "international support."

Following the development of AIA ends and ways, we began focusing on our core task: deriving measures of effectiveness for AIA based on an 8 x 8 ends/ways matrix (see Table S.2). Ideally, we hoped to develop a method and produce measures for every cell in the matrix—e.g., the contribution of professional education and training activities to the ends of access, transformation, and interoperability.

Table S.1
Army International Activities "Ways"

Professional education and training	International support/treaty compliance
Military exercises	Standing forums
Military-to-military exchanges	Materiel transfer and technical training
Military-to-military contacts	Research, Development, Technology, and Engineering (RDT&E) programs

Table S.2
Ends-Ways Matrix

The Ways (from AIAP and TAP)	The Ends (from AIAP, TAP, DPG, QDR, and NSS)								
	Ensure Access	Promote Transformation	Improve Interoperability	Improve Defense Capabilities	Promote Stability and Democracy	Assure Allies	Improve Non-Military Cooperation	Establish Relations
Education and training								
Exercises								
Exchanges								
Military-to-military contacts			Measures of Effectiveness (MOEs)					
International support								
Forums								
FMS + technical training								
RDT&E programs								

To develop the method, we reviewed relevant economics and behavioral sciences literatures and concluded that security cooperation is based on two types of relationships between countries: exchange and socialization. Exchanges are quid pro quo interactions operating mostly in the near term, usually at the program or activity level. They appeal to a target country's self-interest and are generally quantifiable. Socialization operates mainly over the long term and is visible largely at the regional, theater, or HQDA level. It focuses on changing a target country's idea of its national interest. Socialization may flow from repeated exchanges. It generally denotes a qualitative change in a country's attitude or behavior that is often not amenable to quantitative measurement.

Making use of our theoretical work, we also looked for two specific kinds of indicators: output and outcome. Output indicators are the immediate products of AIA. They are usually the products of exchanges that improve a foreign country's ties with the United States in the near term. They are immediate results that can be counted. Examples include number of graduates of U.S. security assistance training programs, senior officer visits, and scientific and technical exchanges. Output indicators lead to measures of performance (MOPs). Outcome indicators are often the by-products of prior outputs. They tend to be more qualitative in nature. They are usually derived from a socialization process that involves building trust and changing foreign perceptions of the utility of working together with the United States over the long term. Outcome indicators are closer to the ultimate goals of AIA and include new capabilities, knowledge, relationships, and standards. They help produce measures of effectiveness (MOEs).

We have developed an extensive set of proposed output and outcome indicators that we have been sharing with AIA officials at the program and command levels and modifying as we gain greater understanding of various programs.

Another task we have performed in cooperation with HQDA and their information technology contractor, COMPEX, is the development of a web-based tool that can be used to solicit AIA program and assessment data from the field. Beginning in the fall of

2004, the Army planned to employ this data collection and reporting tool—AIAKSS (Army International Activities Knowledge Sharing System)—to improve the AIA community's knowledge of the variety of AIA programs and activities as well as to support an ongoing dialogue on how to improve the execution of AIA.

Our final task was to test our assessment concept and data collection and reporting method with AIA officials in three very different organizations: the Army Medical Department, the National Guard Bureau's State Partnership Program, and U.S. Army South.

Key features that distinguish our assessment approach from those that have been developed, or are under development, within the Combatant Commands, OSD, and elsewhere are as follows:

High-Level, U.S.-Centered Focus

Before publication of the OSD Security Cooperation Guidance, the primary objects of U.S. security cooperation policy and analysis were the Combatant Commands and the countries with which the U.S. military has engaged in security cooperation activities. To a certain extent, this resulted in a situation where relatively short-term regional perspectives and foreign interests superseded longer-term global perspectives and U.S. national interests. Our conceptual framework can be used to assess AIA programs around the world according to longer-term goals and criteria approved by both the Army and OSD.

A Solid Conceptual Foundation

Our survey of the relevant social science and management literature—and talks with numerous AIA officials—indicated that exchange and socialization were the ultimate motives for security cooperation. Furthermore, these two concepts were linked in a sequential manner. Quid pro quo exchanges tended to be important early in a security relationship, whereas socialization became a more significant factor in a relationship over time as the number and breadth of international exchanges increased, facilitating a convergence of U.S. and foreign national interests. This theoretical understanding of the role of security cooperation assisted us in developing our AIA output and outcome indicators.

Emphasis on AIA Outcomes

Most systems of measurement used by both the private and the public sectors focus on the immediate results of particular processes: what we call "output" measures. They tend not to focus on the long-term results of programs—i.e., "outcomes"—because outputs are easier to specify and quantify than outcomes and they provide quicker answers to the question of how an organization is performing. Clearly, it is not enough to understand outputs alone if one wants to determine the full effect of AIA on Army and national security goals. Many AIA, such as professional education opportunities for international military students, cannot normally facilitate certain strategic ends, such as improved access to foreign military bases, except over the long term, as U.S.-trained officers rise in rank and gain influence with decision-makers in their countries. This is why we have emphasized the importance of assessing both outputs and outcomes, while working with AIA program/activities managers to develop measures appropriate for their specific programs.

Quantitative and Qualitative Measures

One consequence of our attention to AIA outcomes has been a willingness to consider qualitative, as well as quantitative, assessment techniques. Although quantitative methods have the advantage of succinctness and comparability of data across programs, qualitative measures are often the best or only way to evaluate the long-term effectiveness of particular international programs. For example, knowing the number of students who have passed through the doors of the Marshall Center is not as significant in assessing performance as knowing the ways its alumni networks have been tapped to achieve U.S. strategic goals, e.g., the establishment of stable and democratic civil-military relations.

Reporting Tool Tied to New AIA Database

Other databases and reporting mechanisms exist, or are under development, within the Combatant Commands, defense agencies, and services. However, we agree with HQDA that the Army needs a comprehensive, high-level database to understand and evaluate the full range of international activities the U.S. Army performs. G-3's database and reporting tool, AIAKSS, is not expected to replace existing theater- or program-level databases; rather, it will draw on information collected from these and other sources to provide an aggregated, strategic-level perspective of AIA for decisionmaking on security cooperation at HQDA.

AIAKSS will help AIA personnel to collect and collate data with an unprecedented level of transparency and consistency. It will enable different stakeholders to read the same "sheet of music" when making assessments about the strategic effect of international activities *across the entire Army* rather than within a program, command, or region. Finally, it will also support discussions among programs, commands, and HQDA for future planning and assessment needs.

For HQDA, the main task in launching and sustaining AIAKSS will be to underscore the high-level or "strategic" focus of the AIA assessment process. Past and current assessment efforts have focused on low-level inputs and outputs. A strategic assessment of the effect of international activities, however, will require evaluating how well AIA inputs and outputs have advanced the Army's overall security cooperation objectives. This emphasis on the outcomes of international activities will likely require a shift in the mindset of some AIA personnel, who have not been asked or trained to document the connection between outputs and outcomes. In addition, AIA officials in HQDA and the Major Army Commands must continuously learn from their use of AIAKSS and make appropriate modifications to the tool in accordance with their analytical and reporting needs.

Acknowledgments

Our research team was fortunate to receive considerable guidance and support from our sponsors in the U.S. Army Office of the Director of Strategy, Plans and Policy (International Affairs). We are especially grateful to Mr. Jack Speedy, BG Kevin Ryan, Mr. Jeff Stefani, Mr. Jim Freeman, COL Richard Grabowski, COL Stephen Wilkins, MAJ Timothy Kane, Mr. Hart Lau, Mr. Mark McDonough, Mr. Howard McIntyre, Mr. Michael Adams, Ms. Elizabeth May, LTC Robert Kubler, MAJ Donald Travis, Mr. Kenneth A. LaPlante, COL Mark Volk, LTC Robert Modarelli, Ms. Mary Grizzard, Ms. Molly Bush, and LTC Ara Manjikian.

The development of the Army International Activities Knowledge Sharing System (AIAKSS) was very much a collaborative effort between the HQDA G-3, RAND, and COMPEX. Thus, we would like to thank Ms. Theresa Headley, Ms. Salymol Thomas, and Mr. Steven Jankowski of COMPEX for the determination and creativity they brought to the AIAKSS project.

The implementation of our international activities assessment test cases would not have been possible without the assistance of many individuals in the Army Medical Department (AMEDD), the National Guard Bureau State Partnership Program (NGB SPP), and U.S. Army South (USARSO). These include Mr. Ken Wade, COL Jose Betancourt, Mr. Ken Knight, COL John Storz, Mr. Steven Lemon, Ms. Pat Pinnix, Mr. Sheldon Shealer, Ms. Nancy Crampton, Ms. P. J. Showe, Ms. Judy Williams, LTC Bradford W. Hildabrand, COL Scott R. Severin, Mr. Ivan Speights, Sr., Ms. Vicki Connolly, LTC Brad Hildebrand, and Mr. Ernest Shimada (AMEDD); COL

Mark Kalber, MAJ Michael Braun, MAJ Joe Miller, COL Mike Temme, MAJ Neil Glad, MAJ Paul Schmutzler, SMSgt Wayne Bradford, LTC Charles Brown, MAJ Michael Nave, MAJ Bruce Protesto, MAJ George Spence, MAJ Rich Sloma, LTC Mark Switzer, LTC Dave Thomas, CW4 James Vanas, and CPT Rustin Wonn (NGB SPP); and COL Richard Driver and LTC Al Wood (USARSO).

Our research team would also like express thanks to the following individuals for taking the time to explain their security cooperation programs and provide feedback to us on our proposed assessment framework: COL Thomas Dresen, Mr. Hank Themak, and Mr. William Barr (Office of the Assistant Secretary of the Army, Acquisition, Logistics and Technology); Mr. James Hendrick (Office of the Assistant Secretary of Defense for Command, Control, Communications, and Intelligence); Lt. Col. Robert E. Hayhurst (Office of the Deputy Assistant Secretary of Defense, Force Health Protection and Readiness); Ms. Anna Edmondson (George C. Marshall European Center for Security Studies); COL Michael J. Baier and Mr. Daniel J. Hartmann (U.S. Army Foreign Liaison Directorate); Mr. Dale O. Jackson (U.S. Army Corps of Engineers); Mr. Maxwell Alston (U.S. Army Civil-Military Emergency Planning); Mr. James M. Pahris (U.S. Army Personnel Exchange Program); COL Michael Baier (U.S. Army Headquarters, G-2, Foreign Intelligence Directorate); Mr. Gary Bateman and LTC J. C. Valle (Strategy Division, Plans and Policy Directorate, U.S. European Command); Ms. Lucy Miller, John R. Deni, J. B. Leedy, and Ms. Christina Dall (International Policy Division, U.S. Army Europe); LTC Jay Rudd, Martin Poffenberger, and David Buzzell (U.S. Army Central Command); Ms. Freda Lodge and Ms. Dawn Burke (Defense Security Cooperation Agency); Dr. Andrew J. Corcoran and Dr. Peter Young (United Kingdom Defence Science Technology Laboratory); and Scott Anderson (Calibre Systems).

Finally, the authors would like to thank our reviewers, Kevin F. McCarthy and Douglas C. Lovelace, Jr., as well as Thomas Szayna, the RAND Arroyo Center Associate Director of the Strategy,

Doctrine, Resources Program, for the many helpful comments and suggested revisions that they proffered with respect to an earlier version of this report.

Acronyms

AAR	After-Action Report
ABCA	American, British, Canadian, and Australian Armies' Standardization Program
AKO	Army Knowledge Online
AIA	Army International Activities
AIAKSS	Army International Activities Knowledge Sharing System
AIAP	Army International Activities Plan
AMEDD	Army Medical Department
ASA(ALT)	Assistant Secretary of the Army for Acquisition, Logistics, and Technology
ATOM	Activity to Objective Mapping
CAA	Conference of the American Armies
CENCOM	Central Command
CNAD	Conference of National Armaments Directors
COCOM	Combatant Command
CSA	Chief of Staff of the Army
DoD	Department of Defense
DoS	Department of State
DPG	Defense Planning Guidance
DSCA	Defense Security Cooperation Agency
EUCOM	European Command

FLO	Foreign Liaison Officer
FMF	Foreign Military Financing
FMP	Foreign Materiel Program
FMS	Foreign Military Sales
FORSCOM	Forces Command
GPRA	Government Performance Responsibilities Act
HQDA	Headquarters, Department of the Army
IMET	International Military Education and Training
ISA	International Support Arrangement
ISO PfP	In-the-Spirit-of Partnership for Peace
JCET	Joint and Combined Exchanges and Training
LATAM COOP	Latin American Cooperation Fund
MACOM	Major Command
MDEP	Management Decision Package
ME	Military Exercises
MEDCOM	Medical Command
MFC	Multinational Force Compatibility
MMC	Military-to-Military Contacts
MMEx	Military-to-Military Exchanges
MOD	Ministry of Defence (UK)
MOE	Measure of Effectiveness
MOP	Measure of Performance
MRMC	Medical Research and Materiel Command
MS	Military Strategy
MSLP	Medical Strategic Leadership Program
MT	Materiel Transfer
NATO	North Atlantic Treaty Organization
NIPRNET	Unclassified but Sensitive Internet Protocol Router Network

NORTHCOM	Northern Command
NSS	National Security Strategy
OMB	Office of Management and Budget
OSD	Office of the Secretary of Defense
OTSG	Office of the Surgeon General
PACOM	Pacific Command
PART	Performance Assessment Reporting Tool
PEG	Program Evaluation Group
PfP	Partnership for Peace
POC	Point of Contact
PPBES	Planning, Programming, Budgeting, and Execution System
QDR	Quadrennial Defense Review
R&D	Research and Development
RDT&E	Research, Development, Technology, and Engineering
SCG	Security Cooperation Guidance
SCSC	Security Cooperation Strategic Concept
SIPRNET	Secret Internet Protocol Router Network
SOUTHCOM	Southern Command
SPG	Strategic Planning Guidance
SPP	State Partnership Program
TAP	The Army Plan
TC	Treaty Compliance
TCA	Traditional Commander's Activities
TEPMIS	Theater Engagement Plan Management Information System
TSCMIS	Theater Security Cooperation Management Information System
TSCP	Theater Security Cooperation Plan

TSCS	Theater Security Cooperation Strategy
TT	Technical Training
USAMMA	U.S. Army Medical Materiel Agency
USAREUR	U.S. Army Europe
USARPAC	U.S. Army Pacific
USARSO	U.S. Army South

CHAPTER ONE

Introduction

During the Cold War, U.S. national security policymakers had a single major objective: to contain the Soviet Union. U.S. Army forces were optimized to deter and, if necessary, defeat the Warsaw Pact adversaries in Central Europe, and Army International Activities (AIA) were focused on furthering this objective through cooperation with allies in the North Atlantic Treaty Organization (NATO). The post–Cold War strategic environment is more complex, however. Today, adversaries are often non-state entities, and operations feature coalitions of the willing, composed of both long-time allies and new partners, with a wide range of military strengths and weaknesses.

Such an environment has required that the Department of Defense (DoD) develop a more flexible and comprehensive security cooperation[1] strategy. The first step in this direction occurred in 1998 when Prioritized Regional Objectives in the Contingency Planning Guidance were expanded into Theater Engagement Plans. The second major step was the publication by the Office of the Secretary of Defense (OSD), in 2003, of the first Security Cooperation Guidance,

---

[1] As employed by officials in the George W. Bush administration, security cooperation includes many, but not all, non-combat interactions between the U.S. Department of Defense and foreign military establishments: e.g., foreign military sales (FMS) and training, senior officer visits, and materiel technical cooperation. The term peacetime engagement, as used in the Clinton administration, was defined more broadly than security cooperation. The purpose of engagement was to shape the security environment, and its missions often included positioning U.S. military forces overseas and humanitarian and peacekeeping operations.

which explicitly recognized the role of the military in shaping the international security situation in ways favorable to U.S. interests.

In recent years, a number of important steps have been taken to improve the planning and management AIA.[2] In particular, the Army International Activities Plan (AIAP), first published in 2002, raised the profile of AIA within the Army, offered strategic guidelines for using AIA to meet service- and national-level requirements, and helped to create a greater degree of coherence and identity within the disparate AIA community.

Still, the Army recognizes the need for a high-level assessment mechanism to allocate AIA resources more efficiently, execute AIA programs more effectively, and highlight the contributions of AIA to the Defense Strategy, the OSD Security Cooperation Guidance, and The Army Plan (TAP). For these reasons, in the fall of 2002, Headquarters, Department of the Army (HQDA) G-3 asked the RAND Arroyo Center to develop a conceptual approach to assessing the value of the Army's non-combat interactions with other militaries. The study was to involve four major tasks:

- Elucidating the objectives or "ends" of AIA;
- Consolidating AIA into a manageable set of categories or "ways";
- Establishing linkages between AIA ends and ways through the development of short- and long-term assessment measures; and
- Designing a reporting tool for collecting measurement data from AIA programs[3] and security cooperation officials.

---

[2] Army International Activities are DoD security cooperation activities implemented by U.S. Army personnel.

[3] Currently, there is no standard definition for an AIA "program." Some programs have a dedicated manager at the HQDA or Major Command (MACOM) level. Other programs are managed within the Army in a decentralized fashion. Programs may be funded solely by the Army or may receive funding from Army or non-Army sources. "Program" and "activity" are terms often used interchangeably within the security cooperation community. We make a distinction between the two in this volume, however, intending that an activity be considered as a constituent element of a program.

The hope is that the new assessment framework, which is presented in this document, will be integrated into future versions of the AIAP. A further hope is that this framework will also serve as the progenitor for a "family" of evaluation systems and that such systems will interconnect the major security cooperation players within DoD and, perhaps, the U.S. government.

Security Cooperation and U.S. Army International Activities

All of the U.S. uniformed services have a role in security cooperation, but the Army receives the lion's share of the resources[4] and has been at the forefront of building military-to-military relationships with global partners. This is in part because most countries have some kind of land force to cooperate with, although many partner countries also have an air force, navy, or some variation of these within a security service (e.g., a maritime component within the border guards).

The U.S. Army engages countries around the world through AIA, a large, umbrella-like collection of training, equipping, and consultative programs with multifaceted goals and purposes, whose execution is overseen within the Army Staff by the G-3 Strategy, Plans, and Policy Directorate, Multinational Strategy and Programs Division, G-35-I (DAMO SSI) and within the Army Secretariat by the Assistant Secretary of the Army for Acquisition, Logistics, and Technology (ASA(ALT)).[5] AIA include courses offered at DoD's regional

---

[4] Because of the complexity of the programming and budgeting process with regard to Army International Activities, there has never been a complete accounting of the resources devoted to AIA. For a recent estimate, see Szayna et al. (2004).

[5] ASA(ALT) oversees U.S. Army-executed Title 22 Security Assistance (FMS, Foreign Military Financing (FMF), International Military Education and Training (IMET), etc.), and cooperative Research and Development (R&D), as well as the Engineers and Scientist Exchange Program, the Foreign Comparative Test Program, among other programs. G-3 DAMO SSI is the overall AIA planner, integrator, and resource manager.

centers,[6] international student programs at U.S. Army schools[7] and language institutes,[8] bi/multilateral military exercises, visits and exchanges, planning events, and other meetings involving U.S. Army officials.

Security cooperation activities executed by the Army range from capabilities-building activities through which training and equipment are provided, often via a bilateral or multilateral exercise, to familiarization activities that are not intended to build capabilities but, rather, to build trust, share information, promote mutual understanding of various issues, and discuss security concerns. Some examples of capabilities-building activities include Special Forces Joint and Combined Exchanges and Training (JCET) exercises, educational courses at DoD's regional centers and other U.S. military schools, International Military Education and Training (IMET), and FMF. Examples of familiarization activities include information exchanges, facilities visits, counterpart visits, and some conferences or seminars that are intended to provide training.

A simple categorization scheme for AIA developed by this project and explained in Chapter Three includes education and training, military exercises, military-to-military exchanges, defense and military contacts, international support and treaty compliance, standing forums, materiel transfer and technology training, and Research, Development, Technology, and Engineering (RDT&E) programs.

---

[6] The regional centers are the Marshall Center (Garmisch, Germany), the Asia-Pacific Center for Strategic Studies (Honolulu, Hawaii), the Near East and South Asia Center for Strategic Studies (Washington, D.C.), the Africa Center for Strategic Studies (Washington, D.C.), and the Center for Hemispheric Defense Studies (Washington, D.C.).

[7] Such as the U.S. Army War College in Carlisle, Pennsylvania.

[8] Such as the Defense Language Institute in Monterrey, California.

Overview of U.S. Government Security Cooperation Planning

Army International Activities are planned and executed as part of a larger process whereby guidance is provided by policymakers and operationalized by program and activity managers in the Combatant Commands, the Component Commands, and the services.[9]

At the highest level, the U.S. Security Cooperation Strategy is derived from several key documents. Some of these come from the White House, e.g., the National Security Strategy (NSS) and periodic Executive Orders and functional National Campaign Plans. Others come from the Department of Defense: the Military Strategy (MS), Quadrennial Defense Review (QDR), Strategic Planning Guidance (SPG),[10] and the OSD Security Cooperation Guidance (SCG).[11] These documents are discussed in further detail in Chapter Two. The SCG in particular is now the capstone document for security cooperation. It incorporates information contained in the other key strategic documents.

Using the guidance provided through these key documents, DoD program/activity managers on the execution side then develop regional and country-specific plans to implement the provisions of the guidance. This is a relatively new process that is still being worked out. In the past, country-specific plans, called either the Defense or the Military Plan,[12] were more or less a listing of activities to be conducted during the coming year. Now, these country plans are more strategic; they include goals, objectives, activities, benchmarks for success, and resources. For all Combatant Commands, operationalization of the guidance is found in their regional Theater Security Co-

---

[9] For a more detailed description of the security cooperation planning process, see Szayna et al. (2004).

[10] Before 2004, this document was known as the Defense Planning Guidance (DPG).

[11] Classification levels vary. The NSS, Executive Orders, MS, and the QDR are generally not classified, whereas the National Campaign Plans, SPG, and SCG are classified at the SECRET level.

[12] Joint Staff had country-specific Military Plans and OSD had Defense Plans.

operation Strategy (TSCS), where specific activities and resources are aligned with DoD regional and country-specific objectives. Country-specific Campaign Plans are developed by the Joint Staff and the Combatant Commands. For the services, specifically the Army, the guidance is operationalized in several planning documents, including TAP and the AIAP (see Figure 1.1). The AIAP is analogous to OSD's SCG and is influenced by, as well as acts as an input to, TAP.

Security cooperation officials within DoD make a concerted effort to link the guidance documents as closely as possible to their country plans in an effort to streamline activities, maximize program effectiveness, and minimize confusion. In practice, however, this is no easy task, since those on the implementation side often have multiple masters. Problems also arise when, for example, priority countries

Figure 1.1
AIA in Context

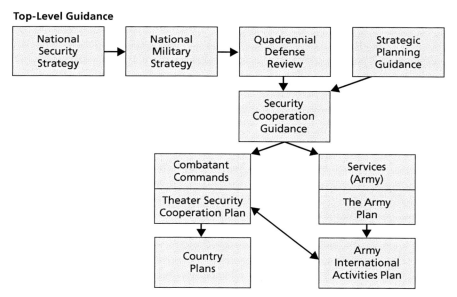

requiring emphasis do not match up in the various strategy documents, making it difficult for implementers to know exactly where they should spend allotted resources. Moreover, in practice, chains of command are sometimes blurred, and personalities, as well as rapid job turnover rates from rotation, play an important role.

The current AIA assessment system is both complex and underdeveloped. Activity-reporting requirements are not institutionalized, and if they do exist, they tend to be stovepiped into the agency (or agencies) that provides the funding, has programmatic oversight, or has country/regional authority (see Figure 1.2). Army Functional Commands report to DAMO G-3 SSI on their non-security assistance AIA programs and to ASA(ALT) and the Defense Security Co-operation Agency (DSCA) on their security assistance programs. Within the regional Combatant Commands, country teams provide defense assessments of the activities within their purview, which they

Figure 1.2
Current AIA Reporting System

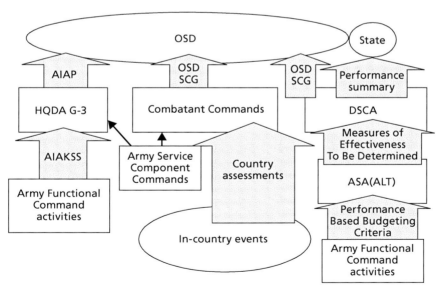

RAND MG329-1.2

provide to the Service Components and the Combatant Commands. For their part, Army Service Component Commands provide programmatic and country-level assessments to HQDA, the Combatant Commands, and DSCA. At present, there is no mechanism for providing AIA (or more generally, security cooperation) assessments to OSD and the Department of State, although OSD's Security Cooperation Guidance is calling for such a system to be established.

At the other end of the chain of command, AIA program managers often work in a vacuum, without full visibility into what other DoD agencies and offices are doing.

System for Measuring AIA Effectiveness

To maximize impact, avoid duplication of effort, and allocate limited resources, a rationale needs to be developed that explains how individual Army International Activities support strategic-level guidance. Moreover, the Army needs a system that allows policymakers, program managers, and implementers to make better decisions about whether ongoing activities should continue, cease, or change in some way. Although the Army International Activities Plan links AIA to larger national goals and guiding documents, the Army does not yet have a detailed, accessible, and adaptable tool for measuring whether International Activities are achieving the objectives identified by the AIAP. A rigorous evaluative framework for AIA—linked to an automated Army International Activities Knowledge Sharing System (AIAKSS)—would help program managers to allocate resources more effectively and assist Army and DoD policymakers in gaining a better understanding of AIA contributions to the National Security Strategy, Military Strategy, and OSD Security Cooperation Guidance objectives.[13]

---

[13] The conceptual framework proposed in this study builds on previous research conducted for the Army on methods for allocating resources to international activities in a more cost effective manner. See, for example, Szayna et al. (2001).

To develop the rigorous framework just called for involves solving a number of problems. We have mentioned some of them above. A full checklist of such challenges would include: (1) the problem created by different stakeholders, with varying responsibilities for management and funding; (2) the issues raised by multiple objectives—of different types (e.g., short- versus long-term) and for different constituencies (e.g., political, diplomatic, and military); (3) the problem raised by a diversity of programs, which makes cross-program comparisons difficult; (4) the issue of causation, which involves determining whether particular programs, as opposed to other factors, actually produce intended effects; (5) the problem of "buy-in," which includes getting the wide variety of individuals responsible for implementing AIA programs to adopt an evaluation system that may not make their jobs easier; and (6) the measurement challenge, given that the achievement of some objectives (e.g., greater U.S. "access" to target countries) could prove difficult to measure quantitatively or even qualitatively.

Organization of the Document

This document is divided into seven chapters and one appendix. Following the introduction, Chapter Two reviews the literature on performance measurement in the public sector, surveys ongoing efforts to measure and assess the performance of security cooperation programs, and identifies the key features of our approach to AIA assessment. This chapter distinguishes between outputs and outcomes in efforts to measure performance and suggests that defining the outcome desired for any given AIA is key to the overall assessment process. Chapter Three explains our derivation of particular AIA objectives or "ends" from U.S. government, DoD, and Army documents. It also provides an explanation for various AIA categories or generic "ways" to achieve AIA ends. In the process, this chapter addresses and responds to the second and third challenges listed above. Chapter Four describes the process we have developed for linking AIA ends and ways. That process starts by defining the logic, including key theories, underlying the concept of security cooperation and then describes the steps we have taken to create AIA performance indicators and assessment measures. In this chapter, we present the remaining elements of our core methodology and explain how we propose to deal (i.e., interactively, for the most part) with the other four challenges listed above: the differences among AIA stakeholders, the issue of causation, the problem of buy-in, and the measurement challenge.

Chapter Five outlines the development of the AIAKSS—a web-based tool for collecting and reporting AIA information that is being made available to international program and command-level officials via Army Knowledge Online. In Chapter Six, we show the results of tests of our AIA assessment approach with officials at the National Guard Bureau State Partnership Program, the Army Medical Department, and U.S. Army South. Chapter Seven describes lessons learned from our AIA assessment effort, potential ways to employ AIAKSS, and some obstacles to its full and effective employment. Finally, the appendix provides a complete listing of the performance indicators we developed in cooperation with AIA programs and security cooperation officials in HQDA G-3 and various Army commands.[14]

---

[14] The performance indicator listing in the appendix is a refinement of the listing provided in Annex D of the Army International Activities Plan (AIAP), Fiscal Years 2007–2008. The indicators in the appendix reflect what we learned from test cases conducted to investigate practical issues in implementing our assessment framework. The listing in Annex D of the AIAP is an earlier version prepared before completion of our test cases in the fall of 2004. The AIAP version of the indicator list is currently incorporated in the AIAKSS.

CHAPTER TWO

Measuring the Performance of Government Programs

Performance measurement has a long history in the private sector as business enterprises have striven to increase productivity and market share.[1] More recently, through the 1980s and 1990s, performance measurement has become an issue in the public sector as a result of public concern for greater governmental efficiency, effectiveness, and accountability. This chapter first describes the performance measurement requirements that have been imposed on government agencies in the last decade and the purposes for such performance measurement. Next, the chapter discusses how performance measurement has been applied to security cooperation. Finally, the chapter outlines the principles of assessment that we derived from our review of the performance management literature and from our discussions with U.S. and allied officials involved in the evaluation of government programs.

Federal Measurement Requirements

The Government Performance Responsibilities Act (GPRA), passed by the U.S. Congress in 1993, requires that federal agencies identify their goals, measure performance, and report on the degree to which

---

[1] See, for example, Bourne and Neeley (2002, pp. 30–31).

those goals are met. Departments and agencies are expected to produce multiyear strategic plans, annual performance plans, and annual performance reports to publicly demonstrate how well they are doing.[2] Each year the Office of Management and Budget (OMB) reviews the performance of a number of federal programs, including those within the DoD. In an effort to help federal agencies, OMB introduced the Performance Assessment Reporting Tool (PART) to evaluate and tie program performance to budget appropriations.[3] PART does not assess the effectiveness of federal programs as much as it gauges how much assessment is built into a program.

GPRA's requirements reflect a shift from performance measurement based on short-term outputs to performance assessment based on long-term outcomes. This means that a program's performance is measured by gains in public safety, program responsiveness, employability of program trainees, and other results that support the goals and objectives of the program and the mission of the agency—rather than, for example, the number and amount of grants made, reports produced, or people trained.

Purposes of Performance Measurement

Although GPRA mandates performance measurement for federal programs, governmental agencies and departments adopt performance measurement methods for a number of reasons. Performance measurement helps an organization to

- Set and adjust goals and standards;
- Detect and correct problems;
- Manage, describe, and improve processes;
- Document accomplishments;
- Provide information to gauge the effectiveness and efficiency of programs, processes, and people;

---

[2] U.S. General Accounting Office (1996).

[3] See information on PART at www.whitehouse.gov/omb.

- Determine whether organizations are fulfilling their vision and meeting their strategic goals;
- Provide measurable results to demonstrate progress toward goals and objectives; and
- Determine the effectiveness of a specific group, department, division, and the organization as a whole.[4]

In short, performance measurement functions as a strategic management tool by demonstrating how well individual programs within an organization are doing, whether they are meeting their goals, whether the users of its products and services are satisfied, and whether improvements are necessary.

Applying Performance Measurement to Security Cooperation

Applying performance measurement to security cooperation is relatively new. However, many efforts are under way in the security cooperation community to gather data at the event level to gauge the effectiveness of international activities and plan for the future. Within the U.S. Army and some regional Combatant Commands there is considerable interest in employing higher-level performance measurement and assessment methods to improve the overall effectiveness of security cooperation.[5] The following is a brief description of the current state of security cooperation assessment in the DoD.[6]

---

[4] Performance-Based Management Special Interest Group (1997, 2001).

[5] Some non-DoD agencies within the U.S. government are also considering the use of performance measurement for their international activities. For example, the Department of State, which has won several national awards for its performance plan, announced in July 2004 that it was applying performance measurement to all its programs. See U.S. Department of State (2003).

[6] The government of the United Kingdom has several studies under way to devise measures of effectiveness for its international programs. The Ministry of Defence is attempting to evaluate its programs using both quantitative and qualitative methods. In addition, the Ministry of Foreign Affairs is working to develop measures of effectiveness (MOEs) for its de-

Major Army Commands

Most major Army commands, such as the National Guard and Army Medical Command, maintain information on security cooperation events and requirements to support organizational planning and reporting. However, data are collected mainly for individual organizational needs, are often not easily accessible to outside agencies, and cannot answer questions on the contribution of AIA to high-level Army and national security goals. Most Army Component Commands, such as U.S. Army Europe (USAREUR) and U.S. Army Pacific (USARPAC), are ahead of their functional counterparts (e.g., the National Guard and the Army Medical Command) with respect to assessment because the former's planning is closely aligned with the TSCS promulgated by the Combatant Commands. For example, international policy planners at USAREUR have developed a strategic concept for security cooperation that guides USAREUR's security cooperation activities and provides specific indicators for achieving desired outcomes in specific countries.

Regional Combatant Commands

Combatant Commanders are required by the DoD to produce a Security Cooperation Strategic Concept (SCSC) document that summarizes their significant security cooperation initiatives. This document must also contain a detailed assessment of their Theater Security Cooperation Plans (TSCPs) effectiveness and a prioritized list of significant resource shortfalls. Although they focus on different issues, the Combatant Commands follow a similar assessment template. For example, the U.S. European Command (EUCOM), U.S. Pacific Command (PACOM), and U.S. Southern Command (SOUTHCOM) SCSC documents begin by describing pertinent national directives, as required by Annex A of the Office of the Secretary of Defense's Security Cooperation Guidance, and then list the Combatant Commanders' theater priorities and associated strategic goals. Next, the documents define end states and objectives and link

---

fense and international cooperation activities. See United Kingdom National Audit Office (2001).

them to specific security cooperation activities in countries within the respective theaters.

Most Combatant Commands actively collect data on security cooperation events and activities in their region in an effort to provide a single integrated operating picture for security cooperation planning and management within their region. Each regional command operates its own database, which is tailored to support its particular needs. Data are collected and stored in the Theater Engagement Plan Management Information System (TEPMIS) in EUCOM; PACOM and SOUTHCOM call their data system the Theater Security Cooperation Management Information System (TSCMIS).

Combatant Commands use various methods for assessing security cooperation activities. Most seek to capture the effects of such activities on countries or regions in their area of interest. For example, in EUCOM, security cooperation officials employ the Activity to Objective Mapping (ATOM) method to grade the performance of international activities in a foreign country on an A to F scale. PACOM assesses country-specific end states and objectives on a quarterly basis, using a traffic light grading system. PACOM also employs models to link user-defined end states and objectives with probabilities that demonstrate the level of influence a particular objective has on a given end state. The results of these assessment tools are incorporated into the Theater Support Command planning process.

Principles of AIA Assessment

In spite of the considerable interest shown within and outside the U.S. Army in performance measurement, current initiatives in this area have not yet produced the kind of information that HQDA needs to assess the overall value of its international activities. Nevertheless, our investigation of the performance management literature, and discussions with officials involved in the evaluation of government programs, supplied us with several basic principles that guided the development of our AIA assessment framework. The following subsections describe these principles.

Measure Effectiveness as Well as Performance

Performance measurement is commonly defined as "the process of quantifying the efficiency and effectiveness of past actions through acquisition, collation, sorting, analysis, interpretation and dissemination of appropriate data."[7] Such a process is carried out to evaluate how well organizations are achieving predetermined goals and the value they provide to stakeholders in accomplishing these goals.[8] Businesses aim to increase productivity and profits by improving the efficiency through which certain outputs are produced. Public programs, by contrast, seek to increase safety, security, welfare, etc. Efficiency is not a sufficient measure of the achievement of these goals. Thus, performance measurement for public programs focuses on measuring effectiveness, that is, how well the products of activities are used to create results that support the goals of the program and organization.[9]

Emphasize Quality over Quantity in Developing Measures

Although accurate assessment relies on a mix of quantitative and qualitative methods,[10] qualitative measures are often better than quantitative measures for evaluating the effectiveness of public sector programs. Numerical indicators, such as percentage increases or decreases, can reveal overall patterns of effectiveness.[11] These quantitative measures, however, are often not any more reliable or informative than qualitative measures in demonstrating how well an activity contributes to a particular goal. Qualitative evaluation methods are better at representing the depth and breath of a relationship. They can specify desired outcomes by focusing on *change*—in circumstances, status, level of functioning, behavior, attitudes, knowledge, skills,

---

[7] Neely (1998).

[8] Moullin (2002).

[9] Derived from Performance-Based Management Special Interest Group (1997).

[10] Lee (1999).

[11] Patton (2002, p. 151).

maintenance, or prevention.[12] In addition, qualitative evaluations can better account for differences in outcomes for activities in different locations. They can help separate a program's effect from the effect of factors external to it: e.g., laws, regulations, politics, and divergent stakeholder interests. Finally, qualitative methods are more appropriate for studying processes that are fluid and dynamic and cannot be fairly summarized on a single rating scale at one point in time.[13]

Establish Linkages Between Outputs, Outcomes, and Objectives

Measuring the effectiveness of government programs requires a deep understanding of how short-term outputs relate to long-term outcomes and how these outcomes are connected to an organization's overall objectives. Such understanding, while generally present in the minds of those who are involved in the programs, is not always documented in the form of data collected as part of regular bureaucratic processes. Even if it is, the information may reside in different locations, since the responsibilities for program planning and execution in the public sector are often split among a variety of individuals and agencies. This means that a concerted effort must be made to take into account the perspectives of a variety of officials regarding the objectives of a program before attempting to promulgate specific output and outcome measures.

Obtain Buy-In from Program Officials

Gaining the trust and approval of program officials is very important to the success of performance measurement. The perception, or suspicion, of control and punishment will inspire resistance from those who are in positions to supply assessment information.[14] This is especially true in the early stages of the process when officials are unaware of the what, why, and how of performance measurement. Thus, developing measures and determining targets is best treated as a multi-

---

[12] Patton (1997, p. 159).

[13] Patton (2002, p. 159).

[14] Pollanen and Young (2001, pp. 10–11).

level negotiation process rather than as a top-down regulatory process, so that program officials need not fear that they will be "asked to take responsibility for, and be judged on, something over which they have little control."[15]

Continue to Refine the Performance Measurement Process

The General Accounting Office has highlighted numerous challenges to developing output and outcome measures for government programs:

- Translating general, long-term strategic goals into more specific, annual performance goals and objectives;
- Distinguishing between outputs and outcomes;
- Specifying how the program's operations will produce the desired outputs and outcomes;
- Getting beyond program outputs to developing outcome measures;
- Specifying quantifiable, readily measurable performance indicators that may not show up for several years;
- Developing interim or alternative measures for program effects that may not show up for several years;
- Estimating a reasonable level for expected performance;
- Defining common, national performance measures for decentralized programs;
- Ascertaining the accuracy of and quality of performance data; and
- Separating the program's effect from the effect of factors external to it.[16]

Because of these potential obstacles, performance measurement must be viewed as a process in need of monitoring and refinement.

---

[15] Patton (1997, p. 158). See also Fitzgerald and Bringham (2001, p. 22).

[16] U.S. General Accounting Office (1997).

As Figure 2.1 indicates, once the initial measurement framework has been established and data have been collected, responsible officials should periodically review the process, forward and backward, to address any gaps and disparities that may have arisen. Looking forward, they should ask themselves whether each step appears to contribute to the following one. For example, do inputs support activities and, if not, why not? What obstacles exist? Looking backward, officials need to ask whether the program's end has been achieved? If not, are they measuring the end correctly? And if so, what are the impediments to achieving the end?

Recognize the Limits of Performance Measurement

Our review of the literature indicates that performance measurement is not an easy process to conduct properly, and there are certain things that it does not or cannot do. For example, because the causes and the effects of outcomes are not easily established, performance measurement often cannot explain why an organization is performing at a certain level or specify what is needed to improve its functioning. In addition, performance measurement does not automatically ensure

Figure 2.1
Performance Measurement Process

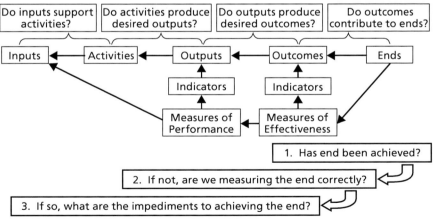

RAND *MG329-2.1*

compliance with laws and regulations. Furthermore, it cannot be used to rank and rate persons or programs.[17] Despite such limitations, however, performance measurement can improve program management and operations by revealing process gaps and inefficiencies and highlighting opportunities to achieve organizational goals with available resources.

The Way Forward

With the foregoing principles in mind, we set out to help HQDA devise its own security cooperation measurement system. We envisioned that such a system would include appropriate AIA ends, ways of achieving these ends, and a method for linking ways to ends. Ideally, it would accommodate a wide range of activities and provide meaningful decision support to headquarters and field-level officials. However, it would not inhibit the execution of international activities. Neither would it create the wrong incentives, incur undue costs, nor cause substantial stress for AIA managers. In the end, the hope is that an AIA assessment system will be able to help HQDA rank and rate—i.e., measure the effectiveness of—security cooperation programs and activities.

---

[17] Performance measurement provides only an approximation of the actual system because metrics are chosen to assess performance toward certain goals, and not all the goals of an organization apply to every level at which an entity operates. Such selective views, therefore, are insufficient to justify any overall ranking or rating of persons or programs. Worse, poorly graded entities might lose resources to do what they do well to support other organizational goals that are important but not central to the specific goals in a particular performance measurement. See U.S. Department of Energy (1996).

CHAPTER THREE

AIA Ends and Ways

The initial step toward developing an AIA assessment system is to derive a workable set of AIA objectives (or "ends") and manageable categories of AIA (or "ways" for pursuing the "ends").[1] In this chapter, we examine the broad range of objectives any state might pursue and distill the key elements of security cooperation into five ideal objectives and eight specific ends. We relate the five objectives to official government guidance—in particular, the National Security Strategy, the Quadrennial Defense Review, the Strategic Planning Guidance, the Security Cooperation Guidance, the Army International Activities Plan, and The Army Plan. We then derive our eight specific ends, describe them, and suggest the importance of each.[2] Finally, we outline the principles we used to develop a manageable set of AIA ways and provide a detailed description of each category.

---

[1] We use the terms "ends" and "ways" in this study to identify the specific sets and categories we are proposing because these terms and our use of them here generally comport with the Army's own usage. For example, we do not employ the term "means" as a synonym or substitute for ways, because the Army uses that term to identify the resources it needs to promote or support ends—i.e., to pay for ways to achieve ends.

[2] Although the AIA ends described in this chapter appear to reflect the aims of security co-operation in general, they are somewhat more specific because they are intended to help HQDA evaluate the international activities that it oversees. Similarly, although our eight ends fit with the broader objectives of the SCG and the AIAP, they are more specific than the three objectives found in the AIAP. Our eight ends provide sufficient detail, without losing too much simplicity or parsimony, to describe and then to evaluate the relationship between Army activities and the broader national aims of security cooperation.

A Method for Deriving Objectives

We employ a two-step approach to define the objectives of Army International Activities to systematically derive ones that are both meaningful and measurable. First, we ask, what is the role of security cooperation in international politics? What might countries, in this case the United States acting through the Army, hope to achieve with security cooperation? This discussion develops some general types of objectives and then works forward from these general categories to more specific objectives, or ends. The categories represent ideal types, or analytical categories that attempt to capture all the logically possible objectives of security cooperation. By starting deductively, we strive to consider all the achievable aims of security cooperation, including objectives that official policy documents might have overlooked. In the second step of our approach, we check these general categorical principles against official national guidance on security cooperation.

Some General Principles for AIA Ends

At the most basic level, security cooperation is a strategy that relies on a set of positive incentives that help a nation—in this instance the United States—to establish, to maintain, or to improve its security relations with its allies and partners. Security cooperation, however, is but one instrument of American diplomacy (see Figure 3.1).

Normally, security cooperation requires that the military downplay its traditional mission of using force or threatening to use force. Instead, the military must offer positive incentives, or carrots, to improve security relations. Such positive incentives come in the form of international activities. Army International Activities represent one facet of American security cooperation efforts, but a large and meaningful component nonetheless.[3]

---

[3] Security cooperation is not always conducted by the U.S. Department of Defense. Other government and non-government organizations sponsor and conduct security cooperation activities.

Figure 3.1
AIA Within the Context of U.S. Foreign Policy

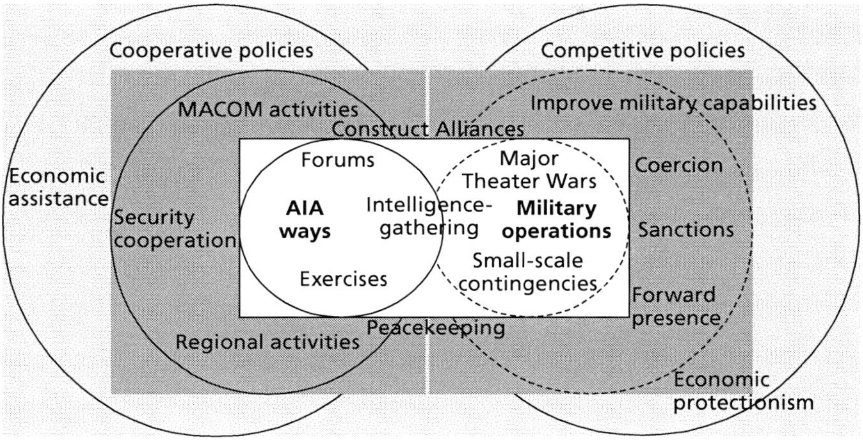

The targets of security cooperation can run the gamut of countries already friendly with the United States to countries with which the U.S. government maintains very few relations. The United States can improve security relations with these allies or partners in a variety of ways. Since the goal of security cooperation is to enhance security relations, international activities will likely endeavor to improve the capabilities of countries, or the assets they use to protect themselves. Similarly, the United States can improve its security relations by engaging in international activities that improve its own military capabilities. Security cooperation can also strive to alter how other countries perceive the intentions of the United States.

These descriptions suggest five ideal types of objectives for security cooperation. First, security cooperation can bolster the capabilities of formal allies or partners. In a sense, this objective seeks to make partner countries self-sufficient in providing for their national defense. Security assistance, or the direct transfer of military equipment or weapon systems, probably represents the most familiar method for improving capabilities. In addition, security cooperation can also seek to bolster the capabilities of allied and partner countries through information exchanges that normally occur during military education or training programs. Military exercises include one more method for improving another country's capabilities. Helping partners and allies improve stability within their borders represents another method of bolstering capabilities.

Second, security cooperation can seek to improve the military capabilities of both the United States and its various allies and regional partners. The most obvious way to accomplish this objective is to promote interoperability between the United States and other friendly states. Standardizing military equipment or agreeing to common military doctrines might represent some additional means for improving coalition military operations. Still further, the states could exchange military technologies or practices that result in improvements for all participating countries.

Third, international activities can seek to increase U.S. military capabilities. From this vantage point, security cooperation strives to increase the capacity of the U.S. military not only to defend itself but also to protect its partners in a particular region. Allies might aid the United States in a conflict by providing access to bases, materiel, or political support for U.S. military operations. Likewise, partners could share crucial military technologies with the United States.

Fourth, security cooperation can aim to change the beliefs and intentions of allies and potential partners. In one sense, security cooperation might try to change how other countries think about their own security practices. These beliefs could include how they view command and control during military operations. For instance, some militaries—especially those trained by the former Soviet Union—have a highly centralized command structure, whereas others, such as those of the United States, employ a more decentralized command structure that emphasizes innovation and adaptation among lower units. These beliefs concern the role of a professional military in a country's domestic politics. Specifically, the United States seeks to impart improved civil-military relations between the governments and armed forces in allied states.

Finally, security cooperation might try to change how countries perceive the intentions of the United States. Here the aim is to reinforce the perception that American foreign and military policy is benign, aimed at creating or maintaining stability across various regions. The objective is to convince countries that the U.S. military is a potential partner for maintaining stability. More concretely, this objective reflects a desire both to reinforce security commitments to allies and to dispel any beliefs that the United States is interested in anything more than promoting stability.

Specific Guidance for AIA Ends

The second step in our approach is to examine how overarching U.S. national security objectives shape the ends of security cooperation for the American government in general and for the U.S. Army in particular. Unlike the previous step, which relies on deductive reasoning, this method assesses how Army International Activities might meet government-wide guidance about military policy. The official policy documents we examined include the National Security Strategy, the Quadrennial Defense Review, the Security Cooperation Guidance, The Army Plan, and the Army International Activities Plan. The first set of documents deals exclusively with the broad objectives of the U.S. government and the Department of Defense. In sequence, they set the parameters for U.S. security policy, working from the presidential level down to the Department of the Army.

This discussion highlights the themes from these documents that are relevant to security cooperation. It is not, as such, intended as a comprehensive review. Any discussion of national security objectives begins with the NSS, a document written by the White House that guides all other types of directives. As the document responsible for laying the foundation for American national security planning, the NSS takes an expansive view of the country's strategic interests. In addition to maintaining stability, the strategy calls for the United States to retain its superior military capabilities to dissuade the emergence of future competitors. The transformation of U.S. security institutions represents one crucial element in keeping this superiority. Still further, the NSS emphasizes the importance of alliances in main-

taining U.S. security. This includes improving existing security relationships as well as establishing new ones.[4]

The Quadrennial Defense Review distills some of these guiding
principles for the purpose of planning at the level of the Department
of Defense.[5] It offers four essential themes. The first two themes deal
with future threats, instructing the armed forces to develop capabilities not only to dissuade potential adversaries from competing with
the United States but also to deter coercive threats against U.S. interests. The third theme of the QDR calls for capabilities that will defeat
aggressive adversaries in the event that deterrence fails. Finally, this
document instructs the U.S. military to assure allies and partners of
U.S. security commitments.

The U.S. Department of Defense Security Cooperation Guidance provides specific instructions that shape the objectives of Army
International Activities.[6] The guidance centers on several themes.
These include the need to combat terrorism, the importance of using
cooperation to assuage regional tensions, the need to strengthen alliances for the future, and the importance of an overseas presence for
the United States. The guidance combines these themes into three
key objectives for security cooperation:

- Build allied and friendly military capabilities for self-defense and
  coalition operations;
- Build defense relationships that promote specified U.S. security
  interests; and
- Provide U.S. forces with peacetime and contingency access and
  en route infrastructure.

---

[4] Bush (2002).

[5] U.S. Department of Defense (2001).

[6] U.S. Department of Defense (April 2003). This document is not available to the general
public. Our study addresses only the unclassified portion of this document.

These three objectives also provide the central aims of the AIAP.[7] As we demonstrate below, our eight ends represent more specific subsets of these three objectives. This relationship makes sense, since our eight ends are derived from such higher-level guidance. Moreover, our list of ends is more specific because we need objectives sufficiently detailed to render them measurable as well as meaningful.

Eight Ends of Army International Activities

Our two-step method suggests at least eight ends for Army International Activities. Although some analysts might argue for more or fewer ends, we settle on eight because that number provides us with enough substantive ends to measure while still maintaining sufficient simplicity to make the creation of measures of effectiveness manageable. We relate each end to our broad discussion of the general aims of security cooperation. Our ends emphasize increasing either American military capabilities, those of partners or allies, or both, or changing the beliefs or intentions of allied or partner nations.

End 1: Ensure Access
International activities serve this end when they improve the likelihood that the U.S. military will receive access to foreign bases, logistical support, or overflight rights. The end calls for access during peacetime as well as during crises. Access permits the United States to project its military power to defend allies and conduct humanitarian operations. Equally important, physical access helps the United States deter potential aggressors in regions where it has important interests.

End 2: Improve Interoperability
International activities advance this end when they improve capabilities of American and allied forces to conduct combined military operations. The end includes improved coalition capabilities across the

---

[7] Headquarters, Department of the Army (2004).

spectrum of military operations, from stability operations to major combat operations. Effective coalition operations are important because they increase the international legitimacy of military action and help the United States share the costs of these operations.

International activities promote interoperability in several ways. In a minimal sense, international activities help countries coordinate military operations with their own services. In addition, international activities improve a country's understanding of how the United States fights, trains, and prepares for military operations. International activities can also increase the military effectiveness of coalition operations by providing opportunities for exercises and vital chances for learning and improving military skills. Finally, international activities can promote interoperability through the sharing of information and technology.

End 3: Improve Non-Military Cooperation

International activities work toward this end when they improve the likelihood that countries cooperate with United States to pursue common interests. This facet of cooperation refers more generally to the support that friendly countries might offer when they cannot participate directly in military operations. Support of this kind can come in the form of sharing the economic costs of operations, political endorsements of American actions, or providing crucial intelligence information. International activities meet this objective by improving American influence in different countries. Ultimately, promoting non-military cooperation sets the stage for deeper cooperative efforts.

End 4: Promote U.S. and Allied Transformation

International activities meet this end when they improve American and allied efforts at transformation. By transformation we refer to efforts to incorporate new technologies and new strategies for warfighting. Security cooperation can facilitate transformation by allowing countries to share new military and information technologies and new ways of warfare. Advancing transformation helps the United States maintain its global predominance in military capabilities.

End 5: Establish New Relationships with Potential Partners

International activities further this end when they serve as an entrée to cultivate new relationship with countries. These are typically states with which the United States lacks any kind of security relationship. U.S. training exercises with Central Asian states represent one example of this kind of objective. Opening the door to cooperation in this way gives the U.S. military an opportunity to establish deeper security relations with these countries. These kinds of interactions pave the way for other American objectives, such as access to foreign military bases or overflight rights.

End 6: Assure Allies of U.S. Commitments

International activities advance this end when they assure allies that the United States will keep its security commitments to them. Countries, especially small states near a potential aggressor, always worry whether their allies will abandon them in a crisis. Because talk is cheap, international activities, like major exercises, provide a costly way to demonstrate a willingness to defend allies. These interactions represent one more opportunity for the United States to show its allies that their security relationship is important.

End 7: Promote Stability and Democracy

International activities meet this end when they help countries improve their domestic security situation. Security cooperation conducted by the American military mostly strives to foster better civil-military relations as one avenue for enhancing the internal stability of allies. Typically this method entails imparting the American tradition of civilian control of the military. In countries where these approaches are not possible or desirable, international activities might help countries defend themselves against criminal elements, insurgencies, or terrorism.

End 8: Improve Defense Capabilities of Allies and Partners

International activities further this end when they improve the military capabilities and self-sufficiency of allies and partner countries. Frequently, foreign military sales are the most direct way to meet this goal. However, military exercises, education, and exchanges might also impart information that improves a country's capabilities. These efforts are important because they provide potential coalition partners for the United States. They also increase the odds that the American military might not need to intervene in a particular region during a crisis because its allies are capable of taking care of the problem.

Are Certain Ends Missing?

Some might argue that this project's set of eight ends omits some important national objectives. For example, our ends do not specifically mention the global war on terrorism. Nor do they reference national concerns over weapons proliferation. From the perspective of our work, such objectives are too specific for Army International Activities. Instead, we derive eight ends that *in combination* can help the United States fight terrorism or halt proliferation. The war on terrorism, for instance, requires assurance, cooperation, interoperability, and promotion of internal stability. Similarly, efforts to halt weapons proliferation depend on assuring allies of American commitments—so they will not acquire their own nuclear arsenals—as well as on improved cooperation with allies.

AIA Ways

There are several reasons for consolidating the large number of Army International Activities into a manageable set of AIA ways. First, developing measures for dozens of individual programs and activities would be too unwieldy and could defeat the purpose of providing an overall AIA evaluation framework for senior leaders. Second, HQDA, the sponsor for this study, has an interest in evaluating the overall progress of AIA, not just individual AIA programs. Third, there is no universally accepted list of AIA,[8] and attempting to establish a defini-

---

[8] The 2005–2006 AIAP identifies 64 programs for which the Army is responsible. However, these programs may include additional subordinate or ancillary activities. A prime example is Multinational Exercises, which is listed as a program but which does not explicitly list the

tive set of such activities could be a long, contentious, and possibly futile endeavor. Finally, AIA categories could eventually be incorporated into the Army's Planning, Programming, Budgeting, and Execution System (PPBES). Currently, AIA are spread across several Program Evaluation Groups (PEGs) and a multitude of Management Decision Packages (MDEPs).

It is challenging to identify a finite set of categories that sufficiently capture the diverse nature of AIA programs. Some programs consist of single recurring events, such as the annual Conference of European Armies. Others are actually organizations that are themselves responsible for a range of activities, the prime example being the George C. Marshall Center for Security Studies. Some programs are bilateral, while others are multilateral, such as regional security exercises. Some involve non-military personnel, such as scientists or defense workers. The programs are also funded through different sources (Department of State, DoD) and overseen by different agencies, ranging from the Army itself to the Department of State. Finally, the Army has varying control over the programs, as some of them are mandated by Congress or required by international treaty.

Organizing Principles Behind the Categorization of Ways

As with AIA ends, it is clear that there is no "right" list of AIA ways. However, our current list does reflect a few guiding principles. First, we have attempted to follow the practices of the larger AIA community. Each regional Combatant Command (COCOM) groups its security cooperation activities into categories provided by OSD's Security Cooperation Guidance. The differences between our categorization and those of the COCOMs are minor and are mainly due to the second principle, which is that the list should be as explicit as possible. For example, whereas the COCOM list has an "other" category, it would not make sense to try to link this category to a national secu-

---

exercises for each of the regional Combatant Commands. Others include the National Guard Bureau's State Partnership Program (SPP) and OSD's Defense and Military Contacts Program, which provides funding to the Army to carry out military-to-military cooperation activities.

rity goal. Thus, our list has a few more categories, such as military-to-military personnel exchanges and military-to-military contacts instead of simply military contacts. The idea behind the current list is that it not only captures the full range of current AIA programs but it will also be able to accommodate the addition of future programs.

The current AIA categorization scheme consists of eight AIA ways:

- Education and training;
- Military-to-military contacts;
- Military-to-military personnel exchanges;
- Standing forums;
- Military exercises;
- RDT&E;
- International support arrangements and treaty compliance; and
- Materiel transfer and technical training.

Below is a more detailed description of each category, including key distinguishing characteristics and examples of each.

Education and Training

This category includes activities that offer professional military education or training for U.S. and foreign military officers or civilians through classroom or field instruction. One or more of the following distinguishes most of these activities: a standardized curriculum, an academic focus, or an academic setting. For example, the IMET program provides training to military officers from allied and friendly nations. The objective of the program is to increase regional stability through effective, mutually beneficial military-to-military relationships that lead to increased defense cooperation between the United States and foreign countries. IMET is overseen and funded by the Department of State and administered by DoD.

Military-to-Military Contacts

This category includes activities that provide interaction among senior military officers, that facilitate decisionmaking between U.S. officers and their foreign counterparts, or that encourage or maintain networks between U.S. and foreign officers. These activities are identified by the focus on senior officers or an emphasis on relationship-building versus formal training. For example, the Chief of Staff of the Army (CSA) Counterpart Visit Program hosts visits to the United States by selected counterparts from key countries, about ten per year. The visits include a ceremonial welcome to HQDA, briefings, and visits to Army installations.

Military-to-Military Personnel Exchanges

This category captures bilateral exchanges of military personnel between the United States and foreign countries. Key characteristics of these activities include a focus on reciprocity between the United States and another country, familiarization versus formalized training/education, or time spent in-country. For example, the Reserve Officers Exchange Program provides U.S., British, and German reserve officers in-country experience, by providing a taste of everyday life in the host nation and through briefings and discussions on NATO issues.

Standing Forums

This category includes programs that plan, coordinate, and implement military standardization as well as facilitate interactions between U.S. and foreign military leaders and subject matter experts. AIA forums focus on the exchange of ideas and are usually well established and multinational in character. For example, the Conference of National Armaments Directors (CNAD) is NATO's highest-level standardization forum. Its primary purpose is to identify and promote opportunities for collaboration in the research, development, and production of military equipment and weapon systems for the armed forces of member countries.

Military Exercises

This category includes bilateral and multilateral military exercises. The key characteristic here is the participation of operational U.S. military units in combat training activities. For example, In-the-

Spirit-of (ISO) Partnership for Peace (PfP) exercises are often directed at developing the capabilities of partner countries to conduct operations with one or more NATO members.

International Support Arrangements and Treaty Compliance
This category captures programs that provide support to other countries, either through official treaties or through humanitarian activities. For example, the Arms Control and Treaty Verification office conducts inspections and multinational visits. The Humanitarian Assistance Program carries out rudimentary construction and renovation projects and provides disaster management training to enhance civil-military operations.

Materiel Transfer and Technical Training
This category includes programs that involve materiel transfer between the United States and foreign countries and any training that accompanies such transfers. In the AIA community, this category captures what is traditionally known as Title 22 Security Assistance programs. The key feature of these programs is a transfer of military goods or contacts, services, and maintenance related to transfers. For example, FMF grants are congressional grants that allow foreign governments to purchase U.S. defense articles, services, and training. FMF may also be used to enhance peacekeeping capabilities, nonproliferation, antiterrorism, or demining programs.

Research, Development, Testing, and Evaluation
This category includes meetings and exchanges of people involved in RDT&E. Activities are distinguished from regular contacts or education by a specific focus on RDT&E issues. For example, the Engineers and Scientists Exchange Program assigns foreign professionals to U.S. defense agencies and contractor facilities for on-the-job RDT&E assignments.

Ends-Ways Matrix

The matrix shown in Table 3.1 represents a first step toward developing a framework for assessing AIA. Organizing AIA ends and ways in this format serves to demonstrate the importance of developing a theoretical rationale for international activities. Before assessing whether AIA programs are effective in achieving an end, one needs to understand the process by which the ways are linked to the ends.

This approach also places the question of effectiveness at a more strategic level of analysis. Rather than asking how IMET facilitates interoperability, it is more useful to ask how educational activities lead to interoperability. It assists the Army in looking at the cumulative effects of AIA, instead of being focused initially on the details of each individual program. This is particularly relevant because many U.S. national security goals may be achieved only over a long time period. Ideally, the framework may assist the Army in learning how to balance its portfolio of AIA by identifying if and where certain types of activities are more effective at reaching certain ends. Potentially, it could even help identify how combinations of different programs influence effectiveness by allowing comparisons across regions and countries. Finally, Table 3.1 provides a template with which to start building actual measures of effectiveness.

Table 3.1
Ends-Ways Matrix

The Ways (from AIAP and TAP)	The Ends (from AIAP, TAP, DPG, QDR, and NSS)							
	Ensure Access	Promote Transfor-mation	Improve Interopera-bility	Improve Defense Capabilities	Promote Stability and Democracy	Assure Allies	Improve Non-Military Cooperation	Establish Relations
Education and training								
Exercises								
Exchanges								
Military-to-military contacts								
International support								
Forums								
FMS + technical training								
RDT&E programs								

Measures of Effectiveness (MOEs)

CHAPTER FOUR

Linking Ways to Ends

This chapter presents our methodology for linking the categories of Army International Activities to the set of national goals that security cooperation logically pursues—i.e., for populating the cells of the matrix depicted in Table 3.1. In the language of the social sciences, such linkages represent hypotheses, or causal processes, that integrate AIA ways and ends. They provide a clear rationale for conducting international activities, help distinguish the short-term benefits of activities from the more long-term objectives, and explain how and why certain categories of activities relate to national security objectives. We took a theoretical, deductive approach toward establishing two types of linkages—exchange and socialization—which we associated with two types of performance indicators—output and outcome. We then used a practical, inductive approach for developing specific AIA output and outcome indicators.

Theoretical Approach to Establishing Linkages

Security cooperation is a process that not only appeals to a target country's national interests but also endeavors to change its national interests. In the short term, AIA will likely appeal to a country's self-interest by creating incentives for cooperation. In the long term, however, security cooperation moves away from merely offering carrots toward building trust and a sense of teamwork between the

United States and the target country. Any fair measure of effectiveness needs to consider these different processes.

Exchange Linkages

Army International Activities promote cooperation and the advancement of American national security policy by appealing to a target country's self-interest through a process of exchange. In return for cooperation, such as access to facilities or political support for U.S. policies, the U.S. Army provides ideas, information, materiel, and technology. This concept of exchange stems from an extensive literature in economics and international relations that views states as self-interested actors facing a set of strategic constraints that influence their decision to either compete or to cooperate with the United States.[1]

Put another way, international activities promote American objectives in three ways: by (a) lowering the costs of cooperation, (b) raising the benefits of cooperation, and (c) signaling future cooperation with the United States.

First, Army International Activities can foster better relations among target countries by lowering the costs of cooperation. Under these circumstances, the U.S. Army exchanges ideas, information, materiel, or technology in return for similar items or in return for cooperative behavior. Another way international activities promote American security policy is by raising the benefits of cooperation. By exchanging some set of goods, the United States obtains a target country's assistance in meeting one of its objectives, such as access to basing rights. Last, international activities can advance American policy by signaling the willingness of the United States to maintain a long-lasting security relationship with a target country. To signal its

---

[1] The concept of exchange and transactions comes from North (1981). Borrowing from non-cooperative game theory used in economics, scholars in international relations rely on a similar logic to explain repeated cooperation among states. See Axelrod (1984).

intentions, the U.S. Army takes a series of costly actions, such as conducting exercises or providing security assistance.[2]

These types of activities are the easiest to assess. Because they entail some form of exchange, they offer clear investments and returns. In addition, these transactions and their potential benefits transpire in the short term. Exchanges get the ball rolling, so to speak, by easing cooperation with the United States and by facilitating future transactions. The process of exchange is easy to see and, therefore, easy to measure.

Socialization Linkages

While exchange appeals to self-interests, socialization attempts to alter how states actually perceive their interests. To use the language of economics, exchange takes a state's interests as given and attempts to make it less costly or more beneficial to cooperate with the United States. Socialization, in contrast, tries to change how states view the costs and benefits of their strategic situation. Rather than seeing the world in terms of "I," socialization strives to help target states see in the world in terms of "We." Socialization achieves this transformation by building trust and creating opportunities for teamwork.

It is important to understand that socialization does not refer to Army International Activities as simple social events, such as dinners or cocktail parties. Socialization instead describes a process where sustained interactions change how countries view the United States and their own security interests. As a growing body of sociological and international relations literature attests, people, firms, and states possess a set of interests that are influenced by their interactions with other actors.[3]

Unlike exchange linkages, however, the process of socialization occurs in the long term, and the benchmarks for gauging progress are not altogether clear. Interestingly enough, assessing when deterrence

---

[2] This discussion relies on Schelling (1960) and Spense (1974). For a more recent treatment, see Fearon (1997, pp. 68–90).

[3] For general discussions, see Tajfel (1981, p. 36), Hogg and Abrams (1998, pp. 31–63), and Goffman (1969). In the field of international relations, the key work is Wendt (1999).

is successful is also unclear. Sometimes deterrence appears to work because a potential aggressor does not attack. An equally plausible explanation is that the potential aggressor never intended attacking. This is the familiar "dogs-that-do-not-bark" problem. Even though deterrence is not easy to measure, however, most analysts tend to think that it is worth pursuing. Given the potential payoffs, namely, sustained cooperation with foreign countries, socialization is also worth the investment.

Socialization can take place in two ways. First, Army International Activities can advance American goals by building trust between foreign countries and the United States. This process occurs over time and begins with exchange mechanisms. Through repeated interactions, a target country comes to perceive not only that American intentions are friendly but also that it can depend on the United States in times of need. A reciprocal relationship develops, as friendly nations show a greater willingness to cooperate with the United States. This type of behavior suggests a key way to gauge the socialization process. Analysts should see that countries with long-standing relationships with the United States are more apt to provide help in times of crisis.

Second, international activities can promote American security policy by building a sense of teamwork among target countries and the United States.[4] Where trust-building emphasizes creating the perception that the United States harbors friendly intentions, creating a sense of teamwork strives to develop the impression that target countries and the American military are interdependent. Put another way, target countries become accustomed to cooperating with the United States. This process not only makes operating together more likely, it also makes it easier.

---

[4] The capacity for groups to form and to act on team preferences is discussed in Sugden (1993, pp. 69–89; 2000, pp. 175–204).

Background Conditions

Security cooperation occurs against the backdrop of international politics. As countries participate in international activities, they also react to their everyday security concerns. In the same way that security cooperation represents only one tool of American foreign policy, target countries must respond to several international as well as domestic demands. Consequently, even when the United States engages in security cooperation with a country over time, exchange and socialization mechanisms crucial to winning cooperation might lose their sway and fail to prove decisive in a particular situation. On occasion, a country may simply fail to cooperate with the United States. Security cooperation, therefore, should probably be viewed as a necessary but not sufficient condition for direct support of U.S. interests.

For example, the American military has a long and enduring relationship with the German and Turkish militaries, but this did not translate into direct military support for the U.S.-led coalition in Iraq in 2003. Instead, Germany and Turkey provided critical, albeit indirect support, to the U.S.-led coalition in Iraq. Such an outcome does not indicate the failure of security cooperation, but it does suggest some important limiting conditions for security cooperation. In the parlance of the social sciences, important variables can confound the effects of security cooperation. Their existence does not mean security cooperation does not matter, but it does mean that there are situations where certain negative factors can thwart its positive effects.

At least two variables or factors can interfere with the positive effects of security cooperation. First, a country might have pressing external security concerns that prevent it from cooperating more deeply with the United States. It may find itself too preoccupied with a threatening neighbor to cooperate with the American military, mostly out of a fear of provoking an external danger. Alternatively, it might have so few external threats that it is not interested in security cooperation. Second, a country's domestic political situation might make cooperation with the United States undesirable. This was likely the case in Germany and Turkey when American diplomats asked for military assistance in Iraq.

In addition to some limiting conditions to security cooperation, other factors can aid exchange and socialization mechanisms. Scholars have isolated three factors in particular that have cultivated deeper relations among individuals, groups, and countries.[5] The first is common fate. When countries believe that they share the same problems, they are more likely to see one another as meaningful partners, thus increasing the likelihood of cooperation. Interdependence is another important factor influencing the development of cooperative relations among countries. When states perceive their international situation as tied together, where the choices of one state affect the choices of most other states, they are more likely to cooperate with one another. Third, when countries perceive that they are alike, in terms of their political beliefs and institutions, they are more likely to see cooperation as beneficial. The perception of political homogeneity, then, reduces the number of conflicts that might otherwise arise among nations.

From Linkages to Indicators

Because exchange and socialization mechanisms influence countries in different ways, the kinds of evidence or indicators that will confirm their presence will also differ. Indicators are important because they allow analysts to identify signposts of security cooperation actually taking place and, ultimately, to measure the effectiveness of international activities. Without linkages, however, analysts would have to rely on guesswork to arrive at the appropriate indicators needed to measure the effectiveness of international activities, or any activity for that matter.

---

[5] For a detailed treatment, see Wendt (1999, pp. 343–366). Wendt also lists self-restraint as another important attribute of socialization. We choose not to include this factor because, in our view, the presence of common fate, interdependence, and homogeneity leads countries to exercise self-restraint.

Exchanges and Outputs

We generally associate the concept of exchange with output indicators. Short-term international activities help the United States to further its security cooperation goals through activities that anticipate a particular response or result. Their purpose is usually to modify a foreign country's ties with the United States in the near term. For example, changing the number of billets offered to a certain country for professional education and training could signal an interest in closer ties or displeasure with certain aspects of the relationship.[6] Examples of output indicators could include graduates of U.S. security assistance training programs, senior officer visits, as well as scientific and technical exchanges.

Socialization and Outcomes

Outcome indicators are often the products of prior outputs. They are usually derived from a socialization process that involves trust-building, networking, and changing foreign perceptions of the utility of working together with the United States over the long term. Outcome indicators are closer to the ultimate ends of AIA and include new capabilities, knowledge, relationships, and standards.

Figure 4.1 illustrates how output and outcome indicators fit within the AIA ways/ends framework by demonstrating the relationship between education and training activities and cooperation between the United States and its allies and partners. The input of resources (money and manpower) enables the conduct of education and training activities. An output of such activities is the number of graduates. Exchange at this point involves the transfer of U.S. resources to enable visiting foreign army officers to acquire the principles of democratic governance and security and to transfer what they have learned to their home countries.

---

[6] Congress recommended a suspension of exchanges with Malaysia under the IMET program in October 2003 in response to remarks made by the Malaysian prime minister.

Figure 4.1
Putting Indicators in Context

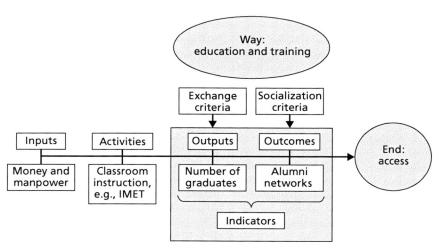

The presence of an alumni network is one outcome indicator in that it provides the Army with a mechanism to reach out to foreign individuals in support of the ultimate end of access. Alumni networks promote socialization by building a support system and professional linkages for graduates to tap into as they work to put into practice what they have learned.

An education and training program might report that it produced 50 graduates in FY 2003 as a measure of its output for that year. An outcome measure for the end of access could be that its alumni network is now five years old and maintains a membership of 500 persons. These output and outcome measures suggest the availability of certain tools, specifically persons, relationships, and institutions that the Army can use to improve the U.S. military's chances to obtain physical access to strategically important countries.

Practical Approach to Developing Indicators

Our survey of the theoretical literature on exchanges and socialization allowed us to think conceptually about measuring security cooperation performance. However, interviews with Army personnel involved in the planning and execution of international activities underscored certain practical issues that must be considered in establishing linkages between AIA ways and ends. Even though we thought of output indicators in terms of exchanges, and outcome indicators in terms of socialization, we recognized that in some circumstances, outputs of activities could lead directly to socializing behavior and changing perceptions about the United States. Also, in some circumstances, a desired outcome might be more of an exchange. We also realized that some outputs can lead directly to the accomplishment of the AIA end, for example, a quid pro quo offer of training in U.S. schools for access to foreign territory. However, in general, security cooperation activities will seldom result directly in the attainment of a national objective (see Figure 4.2). The practical approach to developing indicators thus requires flexibility and imagination in searching for evidence of success.

Guidelines for Identifying Indicators

For the practical matter of identifying indicators for outputs and outcomes, we were mindful of the following:

- Ask for answers, not simply data. Knowing that ten alumni directories exist does not tell us how well they serve to connect people and contribute to the AIA ends. The Army needs to know whether these directories are used, how frequently, by whom and for what purpose, and their significance to accomplishing Army and national objectives.
- Avoid drowning in data; limit to what is meaningful to decisionmaking.
- Do not bias in favor of quantitative data because AIA success can often be best measured by qualitative evaluation.

Figure 4.2
Possible Types of Indicators

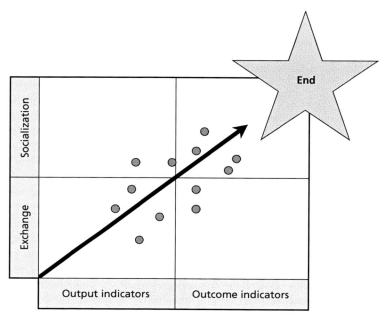

- Be aware of cost of measurement; leverage what is already collected and consider carefully what additional data are needed.
- Note that there is considerable interest within the U.S. Army to improve the effectiveness of AIA through performance measurement, but there is just as much trepidation over the extra work that may be required of people who are already stretched thin when it comes to their time.
- Choose indicators that encourage people to think in terms of performance planning and not performance reviews so that staff energy is not wasted on justifying what has already occurred.
- Ensure that output and outcome indicators are sufficiently broad to be applicable to various AIA; they must also be sufficiently specific to be meaningful in assessing whether outputs and outcomes are contributing to the AIA ends.

Candidate Indicators for Outputs and Outcomes

For our purposes in designing useful indicators, we focused on whether and to what extent the outputs and outcomes of AIA support the Army's desired ends for security cooperation. Although AIA are "international" activities that ideally incur benefits for all participants involved, the emphasis of the measures developed for this exercise is on the benefits for the Army and the United States.

The concept of exchange and socialization linkages discussed above guided us in tying activities to ends, but we also searched for concrete and observable indicators. For example, in the short run, U.S. Army schools will produce more U.S. and foreign Army officers familiar with foreign and U.S. military thinking and institutions, respectively. This can influence a change in perspectives and mindsets as well as expand professional networks. In the longer term, these individuals may rise to positions of higher leadership, where they will be in a position to influence their countries' security relations with the United States, which in turn could include the granting of access rights and logistical support. We methodically repeated this exercise with every category of activity and every AIA end to produce an initial list of possible output and outcome indicators for each AIA end.

In refining the indicators, we began looking for patterns and unifying concepts to characterize these indicators. The result is three classes of indicators, namely, people, things, and costs.

- Indicators to measure people would include:
  — Number of billets
  — Number of graduates
  — Rate of placement
  — Rate of promotion
  — Assignment to high-priority countries
  — Number of lecturers exchanged or sent
  — Presence of formal alumni or professional networks.
- Indicators to measure things would include:
  — Equipment or technology transferred
  — Types of classes (or curriculum)
  — Number of classes

— Doctrinal changes
— Offers of troops, access, logistical support, etc.
— Number of exercises
— Types of exercises
— Numbers of meetings or forums
— Types of meetings and forums
— Type of U.S. assistance
— Types of agreements (treaties, memoranda of understanding, leases, etc.)
— Number of parties involved (bilateral or multilateral)
— Level or rank of parties involved
— Statements of support.
• Indicators to measure costs would include:
— Costs for training
— Costs for management support
— Costs for logistics
— Costs for operations
— Costs for R&D
— Costs for foreign assistance
— Costs for military exercises
— Costs for conferences
— Costs for planning.

As we worked from the framework to the specific indicators and then back from the specific indicators to the generic elements of security cooperation, we began to see patterns of activity and their causes and effects. The easiest way to understand these connections is to think of security cooperation output and outcome indicators as a coherent narrative; that is, these indicators taken together should tell a cogent story of a security cooperation activity and its effect on AIA ends. During the summer of 2003, we presented our indicators to more than 20 Army officials, received feedback, and revised our indicator lists based on their responses. Our proposed indicators provoked many thoughtful discussions concerning the role of AIA in promoting short- and long-term security cooperation objectives. A full list of these indicators arranged by AIA ends can be found in the appendix.

Transforming Indicators into Measures

The next step in the development of our methodology was to transform the generic output and outcome indicators we were producing into specific measures of performance (MOPs) and MOEs. To use the education and training example again, we needed to find a way to translate output and outcome indicators for education and training programs in general into a specifically desired output and outcome for a particular education and training program. Figure 4.3 illustrates the translation of the output of this program into a concrete number of graduates, ten, which is intended to represent a reasonable target that the program can hit for a particular country of interest to

Figure 4.3
Measures of Performance and Effectiveness

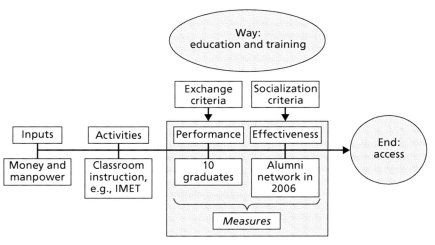

the United States. That number becomes the MOP in this case. Likewise, we have set a specific goal as the desired outcome of this program, namely, the establishment by 2006 of an alumni network comprising these graduates. Whether that goal can be achieved as stipulated becomes, in effect, the MOE. The ultimate goal, of course, is for such an alumni network to contribute directly to one of the "ends"—for example, by helping to facilitate U.S. military "access" to a strategically important foreign country.

Several considerations guided us in this phase of methodological development. We sought to

- Integrate our performance measurement system with performance management.[7] We wanted our system to link planning, implementation, and assessment. Such linkage should enable senior administrators to communicate the organization's mission, vision, values, and strategic direction to employees and external stakeholders. From both employees and stakeholders, the system should also produce "feedback" intelligence that supports senior administrators' efforts to manage the organization's business plan, data systems, and budget processes.
- Build a dynamic process. In our view, selection of specific performance and effectiveness measures should be driven by the high-level AIA ends with full consideration of realities at the activity level. Our system should encourage and enable regular interchanges between both levels.
- Set targets through consultation. Senior administrators at HQDA should maintain a continuing dialogue with those responsible for planning and managing the Army's international activities in the field. Ideally, specific goals to be achieved or targets to be met by these activities should be the product of this dialogue—the result of mutual agreement, as opposed to dictates from on high.

---

[7] For a discussion of an integrated performance management and measurement system, see Performance-Based Management Special Interest Group (2001).

- Link strategic planning to program implementation and evaluation. Once set, preferably by mutual agreement, goals or targets should be revisited to see if they are, in fact, being met. If not, changes may be warranted: in the targets themselves, in other measures, or in the inputs to particular programs (see Figure 2.1).
- Leverage current data collection efforts and support the construction of an AIA database. This database should provide the factual basis to support the continuing dialogue between higher and lower management levels that links planning, implementation, and assessment in our system.

In terms of the education and training example depicted in Figure 4.3, therefore, we would expect HQDA to engage in a dialogue with program or activity managers responsible for this education and training program. Through a process of negotiation that, we hope, produces mutual agreement in the end, as well as buy-in, we envisage the establishment of specific output targets (MOPs) and outcome goals (MOEs) to be met by the program in the future, with all parties relying on the AIA database for current information. The dialogue will continue, and in succeeding years, it will also include assessments of how well the program is performing relative to the MOPs and MOEs originally set for it. If these evaluative reviews point to a need for change—whether in the measures, the indicators, the activities, or the inputs—it can be effected, as well as subsequently reflected in future iterations of the AIA measurement and assessment process we are proposing here.

CHAPTER FIVE

Army International Activities Knowledge Sharing System

With the development of a detailed list of output and outcome indicators linking AIA ways and ends, as well as a methodology for converting these indicators into specific MOPs and MOEs, we completed our conceptual framework for assessing the value of Army International Activities. What remained was to embed this assessment framework within a widely distributed, computerized tool that would permit HQDA and its Component Commands to collect and report the data needed to develop specific MOPs and MOEs. The result was the AIAKSS. Designed to serve all Army personnel involved in the planning, execution, and performance of security cooperation activities, AIAKSS was a collaborative effort between HQDA G-3, RAND, and COMPEX, an information technology firm contracted by the Army. The development of this web-based tool began in November 2003 and continued through the fall of 2004. At the time of the writing of this report, AIAKSS was available for "read-only" purposes on the U.S. Army's Intranet, Army Knowledge Online (AKO). During 2005, the HQDA G-3 intended to use AIAKSS to collect data on AIA programs throughout the Army to meet the assessment objectives established in the Secretary of Defense's Security Cooperation Guidance.

Practical Concerns

In addition to our conceptual assessment methodology, we took practical concerns into consideration when thinking about how to create a computerized database and reporting tool that would provide meaningful programmatic and assessment information to the Army. For example, Army International Activities vary greatly in mission, structure, funding, size, etc., and we recognized that AIAKSS must be built to accommodate this diversity. In most cases, program or higher-level resources will not be available to hire personnel to collect and report data for a new information system devoted to AIA. Also, Army personnel are rotated regularly, and, as a result, most program personnel charged with collecting and submitting AIAKSS data will not have previous experience with the system.

Strategy for Building AIAKSS

Table 5.1 summarizes our strategy for integrating our assessment methodology into a multipurpose security cooperation database and reporting tool, and the practical concerns expressed by AIA officials. The following subsections describe the basic elements of our strategy, our thinking with regard to these elements, and the corresponding features and functions that were built into AIAKSS.

Access

AIAKSS is intended to enable the sharing of information on international activities across the Army while also collecting assessment data for senior Army and DoD leaders. For this reason, the Army decided to place AIAKSS on AKO, the Army's Intranet. Through AIAKSS, all Army personnel will be able to learn about international activities, where they occur, what results they produce, and how they are tied to national and Army strategic goals. Readers will also be provided with

Table 5.1
Summary of AIAKSS Development Strategy

Strategic Elements	What Guided Our Thinking?	What Is Built into AIAKSS?
Access	Enable information-sharing in a protected environment. Ensure data integrity and security.	AIAKSS is placed on Army's Intranet for Army-wide access. Access for data providers is controlled.
Data storage	Make AIAKSS the central repository for information on AIA.	AIAKSS will store all data collected over time in a structured and secure environment.
Data collection	Ask only for data useful to assessment. Leverage existing sources of data. Develop a standardized (albeit flexible) process for data submission for all AIA programs.	Data fields are closely tied to indicators and other information that the Army must use to conduct analysis and assessment. The system architecture includes two tiers of data collection capacity that works for all AIA programs.
Measurement indicators	Develop a set of indicators that are applicable to a wide range of AIA programs.	Generic indicators are provided, but users can propose alternative or additional indicators.
Data collation and searches	Collate data in ways that will support queries in breadth and depth and answer general and specific questions about AIA.	Two-tier system architecture and availability of checklists and drop-down menus support data collection and collation requirements.
Factors impeding success	Allow AIA officials to explain impediments to the performance and effectiveness of their activities and programs.	AIAKSS collects data on challenges encountered by programs.

AIA program contacts in case they require additional information. Program officials can also use AIAKSS to identify counterparts in other commands and regions whose contacts and experience they might use to support the planning, execution, and assessment of the activities for which they are responsible.

Mindful of the need to protect the integrity of AIA data, we recommended to the Army that only AIA program personnel authorized by the AIAKSS administrator be permitted to provide inputs to the system. Password protection will also be instituted so that only Army personnel and special guests can access AIAKSS. In addition, lower-level "event" data are accessible only to personnel responsible for a program and the AIAKSS administrator at HQDA. General users with read-only permission will not have access to these data.

Initially, HQDA wanted to locate AIAKSS on the military's classified Intranet, the SIPRNET, to expand the type of security co-operation information that would be available to users. However, many Army offices in the United States and overseas do not yet have access to the SIPRNET, so this step would have contravened the Army's goal of promoting the use of AIAKSS throughout the service as well as complicated the data submission process for many program officials. Therefore, we suggested placing AIAKSS on the unclassified NIPRNET until most of its potential users had gained access to the SIPRNET.

Data Storage

Currently, there is no central repository of information on Army International Activities. AIA program officials produce reports describing their particular objectives and accomplishments, but these reports come in a variety of formats, are published at different times, and are not readily obtainable by the entire security cooperation community. Establishing a primary locus of AIA information is thus a necessary step if the Army wishes to gain an overall perspective on the strengths and weaknesses of its security cooperation programs. AIAKSS is well suited to perform this function. It provides a set of carefully constructed data fields, a system architecture that permits the storage of disaggregated and aggregated data, and a secure AKO environment.

Data Collection

Taking into account the varying sizes, purposes, and types of AIA programs, we proposed a flexible but uniform process for AIAKSS data submission as well as standard fields to support data collation and analysis. For example, AIA programs and activities inputting data into AIAKSS are required to list a point of contact so that users who have questions about the source or clarity of the data they read on the system know where to go for answers. Requiring authorization for personnel submitting information to AIAKSS will also reduce the likelihood of confusion or corruption of data as a result of overlapping entries. Finally, AIA program officials, generally at the command level, can use AIAKSS to collect data from subordinate or associated activity managers to facilitate the former's reporting to HQDA. This feature is optional and was created to serve program personnel rather than HQDA. This dual-level collection tool is explained further below.

Measurement Indicators

Using the methodology discussed in Chapter Four, we developed an initial set of output and outcome indicators in the summer of 2003 to demonstrate what AIA programs produce and how these "products" contribute to U.S. Army goals. We then refined these initial indicators through three in-depth test cases, described in Chapter Six. These refined indicators, completed in the fall of 2004, are listed in the appendix. As part of the annual AIAP evaluation process, HQDA will ask AIA officials within subordinate commands to choose specific output and outcome indicators from the overall list in AIAKSS, based on the goals and categories of activities that pertain to their programs and activities. Using the same set of generic indicators to support measurement across programs provides the Army with a common yardstick to support analysis and assessment. At the same time, flexibility is built into the system by allowing program officials to nominate alternative (or additional) indicators when the available indicators do not adequately represent a particular program's outputs and outcomes. We have proposed that HQDA G-3 SSI review indicator nominations and modify the AIAKSS indicator list as appropriate.

Data Collation and Data Search

To make the data collected on AIAKSS useful to the entire Army security cooperation community, it must be collated in ways that support general and specific queries concerning Army International Activities. Our determination of the data fields and system architecture for data collection laid the foundation for data collation and searches.

Different search functions were incorporated into AIAKSS to meet different user needs. For example, data can be searched and sorted by year, program goals, type of activity, and regions of the world. In addition, AIAKSS contains automated checklists and drop-down menus to facilitate searches. Users can also save their searches online or export results in various formats for use in briefings and reports.

Factors Impeding Success

Sometimes factors exogenous to programs, such as natural disasters and political instability, can derail an effort or adversely affect its performance. Therefore, it is important to collect information that helps explain AIA performance and effectiveness beyond measurements of their outputs and outcomes. To enable this, AIAKSS is designed to collect data on challenges to the achievement of security cooperation objectives that are largely outside the control of AIA officials. Additional programmatic impediments may include changes in laws and policies, shifts in political relationships and conditions, resource shortfalls, increases in unit operational tempo, communication failures, and information gaps. Program officials have the option to report these exogenous factors and to explain the steps they have taken to overcome them, if any. These "lessons learned" may provide insights to other programs faced with similar problems. Providing this information will also enable HQDA to develop a more comprehensive understanding of program performance and, possibly, suggest ways to assist its subordinate commands in carrying out their AIA responsibilities.

Detailed Description of AIAKSS

This section provides a more concrete description of AIAKSS. It begins with a description of the major AIAKSS users and then explains the system's major functions and structural components.

AIAKSS Users

All Army personnel with access to Army Knowledge Online can access AIAKSS. Further access control permits three types of users: readers, authors, and AIAKSS administrators.

- Readers are all Army personnel and other persons with access to AKO. They can perform searches to generate various types of reports to meet their informational needs. Readers do not need to obtain special approval from the AIAKSS administrator to access the system.
- Authors are Army personnel given responsibility by their commands to enter data into AIAKSS. They can enter new records and revise existing ones. Authors will require a one-time approval from the AIAKSS administrator.
- AIAKSS administrators—currently personnel in HQDA G-3/SSI—control access to the system. They are also responsible for system maintenance, oversight, and answering queries from all users. At this time, only AIAKSS administrators can approve additional output and outcome indicators.

AIAKSS Functions

AIAKSS has three main functions. First, it allows AIA officials to submit data on the performance and effectiveness of their programs and activities. Second, it allows AIA officials to review and edit the records they have created. Third, it allows all AIAKSS users to conduct searches to obtain information about Army security cooperation.

- Create new records. All U.S. Army Major Commands that manage AIA programs are expected annually to provide data on their performance/effectiveness during the previous fiscal year to HQDA through AIAKSS. Army officers designated by their commands to report on their AIA programs will function as AIAKSS "authors" for reporting purposes.
- Review/edit an existing record. Authors can return to the system to continue data entry or make changes to unfinished records.

- Search for data. All AIAKSS users can search for data and generate textual or graphical reports. Searches can be tailored to user needs and queries can be saved on the system for future use. The reports generated can be converted into Rich Text Format and Microsoft Word, PowerPoint, and Excel spreadsheet formats and downloaded to the users' own computers. This enables data to be easily inserted into briefings and other types of documents.

AIAKSS Structure

AIAKSS has two main structural components. The first component involves the collection of data related to international programs/ activities, to include basic descriptions, associated ends and ways, output and outcome indicators, and challenges encountered during program/activity execution. The second component allows users to conduct searches and generate a variety of text and graphical reports to support analysis, assessment, and decisionmaking.

There are two levels for data collection and searches. Although AIAKSS is intended as a tool to support program-level reporting to HQDA, officials at the Major Commands frequently cannot perform this function without first collecting data from subordinate officials, who often take the lead in organizing and executing AIA and often have better awareness of the outputs and outcomes of international activities. Therefore, RAND and COMPEX developed a "field-level" data collection and search capability for AIA officials responsible for ongoing activities below the program level.[1] Activity authors simply select the field-level option when entering data. The main difference between the two collection operations is that AIA program or command personnel determine authorization for authors in subprogram units. Also, activity data are visible only to the program and AIAKSS

---

[1] Information on individual events is currently stored in a variety of databases, including the Theater Security Cooperation Management Information Systems operated by several regional Combatant Commands. Although we do not envision AIAKSS becoming an events database, it may be desirable in the future to create an interface between AIAKSS and existing events databases.

administrator. Army personnel with "read" access in AIAKSS will only be able to read program-level data.

Figure 5.1 shows a simplified view of two possible data aggregation methods that could be used by functional or regional Combatant Commands. In Example A, an Army-funded international program office within a command is responsible for collating information from subordinate activities and entering the data in AIAKSS. In Example B, an AIA official on the command staff is responsible for "rolling up" data on various command programs, including non-Army-funded programs, whose activities are carried out by units subordinate to the command. Suggested strategies on completing the data aggregation process are included in the final section of this chapter.

Data Collection. The data collection mechanism within AIAKSS has three parts: Part A collects basic program information, Part B collects output and outcome data, and Part C collects data on challenges to program success.

Figure 5.1
Dual Data Collection Structures

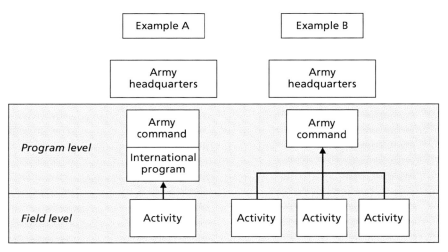

With regard to Part A, basic program information includes the name of the program or activity, a statement of its purpose or mission, its proponent, its authority, and a point of contact. Another basic information category is funding. Programs are required to report their funding sources (e.g., U.S. Army Title 10, Title 22 Security Assistance), level of appropriations, the total amount funded, and total amount of unexpended funds for the year. Programs will also identify the Combatant Commands, regions, and specific countries in which they operate, and select relevant AIA ways and AIA ends for each country. When the scope of activities contained within a program is relatively narrow (e.g., the Fifth U.S. Army's Border Commanders' Conference with Mexico), the choice of a particular way is evident (e.g., standing forums). However, broad-based programs will often require the selection of more than one way. By the same token, the stated mission of a program might motivate the choice of one or more AIA ends.

Part B focuses on the specific outputs and outcomes of Army International Activities. The system automatically generates indicators based on the category of activity and goals chosen in Part A. Authors then select the output and outcome indicators that best represent the achievements of their security cooperation activities.[2] They are then required to provide evidence for their selection of indicators in the form of citations from a report, an official memo, or another type of document. Should the embedded list of indicators fail to adequately represent programmatic achievements, AIA officials can propose alternative or additional indicators. Nominations should be sent via AIAKSS to the system administrator, who will review and approve them as appropriate. New indicators will be incorporated within AIAKSS and thus made available to all program officials. This allows some measure of control over assessment data to promote cross-

---

[2] AIAKSS authors can indicate whether the outcome indicators they select are directly linked to the outputs reported for the current year or result from outputs produced in previous years. This permits AIA officials to analyze the cumulative effect of AIA programs as well as their immediate consequences.

program comparisons as well as a level of flexibility to accommodate differences among security cooperation programs and activities.

Another important component of Part B is the capabilities list in TAP, which is fully reproduced in AIAKSS. After providing information on AIA outputs and outcomes, AIA officials are asked to state TAP capabilities that have benefited or been strengthened as a result of AIA programs. This allows AIA officials to indicate how their programs directly contribute to the execution of the Army's Title 10 responsibilities.

The success of Army International Activities depends on more than good planning and hard work. Part C provides an opportunity for AIA officials to provide feedback to Army and DoD leaders on external factors hindering the progress of security cooperation and shares this information with other Army personnel. Tracking challenges can highlight organizational, doctrinal, or funding issues that HQDA might be able to address. It can also indicate problems over which the United States has little control, which may or may not cause senior officials to reconsider a program's utility, depending on the circumstances. AIAKSS currently lists the following six types of programmatic challenges:

- Statutory: treaties, laws, regulations, or policies that bar or hinder the execution and success of Army International Activities.
- Political: important changes in relations between the United States and a foreign partner country or changes in government or leadership that bar or hinder execution and success of Army International Activities.
- Resources: insufficient resources (money, manpower, materiel) available to support execution and success of Army International Activities.
- Operational and personnel tempo: unit operational tempo is too high to enable personnel to conduct Army International Activities. Personnel are required to shift their focus from security cooperation to operational activities.
- Communication failures and information gaps: technical difficulties, data and knowledge gaps, and other communication

problems that bar or hinder execution and success of Army International Activities.

- Weather and natural disasters: earthquakes, hurricanes, floods, and other forms of natural disaster and severe weather conditions that bar or hinder the execution and success of Army International Activities.

AIA officials can select all the challenges that apply. If no challenges were encountered, programs can skip this section. There is also an "Others" category for unlisted challenges. For each challenge that is chosen, AIA officials must report whether the problem existed in a foreign country or in the United States. They will provide details on the challenge and actions taken. They will also indicate whether the challenge resulted in a cancellation of security cooperation activities and whether activities will continue in the next fiscal year. Finally, AIA officials are asked to share lessons learned from their experience in dealing with the challenge.

Data Search. There are three ways to conduct AIA program searches using AIAKSS. First, a basic search function generates a full program report for the year chosen. All information submitted about that program would appear in a single report. The second is a tailored search method that allows users to choose from a large menu of fields, including year, AIA ends, AIA ways, MACOM, and country. Finally, AIAKSS has a geographical search function. Users simply point to a region on the map and drill down from there.

AIAKSS has a couple of other notable search features. It permits users to save their search criteria for future use. This "advanced" search function keeps users from having to reenter the same query. When a query is repeated, AIAKSS will retrieve the most recent data available. Another search function allows users to find the Point of Contact (POC) for an international program or activity. This feature was included in response to feedback from AIA personnel who wanted a way to easily identify colleagues involved in planning and executing international activities. The POC search will allow personnel to better communicate with and leverage the experience, professional networks, and other assets of the AIA community. Users can create a directory or conduct specific POC searches by name, AIA way, AIA end, program name, COCOM, or other keyword.

AIAKSS Issues

As mentioned above, HQDA G-3 planned to use AIAKSS to collect data from AIA program officials beginning in 2004–2005. This section raises some of the issues that HQDA G-3 will need to address in 2005 and beyond if AIAKSS is to reach its full potential as a comprehensive AIA database and high-level assessment tool.

Which Programs Should Be Assessed?

Should the G-3 focus on programs that are funded directly by the Army, using the services own Title 10 resources? If so, that would exclude Title 22 security assistance programs managed by the Army but funded by foreign governments and the Department of State. It would also exclude OSD programs, such as the Warsaw Initiative Fund, that support many international activities (e.g., multinational exercises) executed by the Army's Component Commands in the regional theaters. On the other hand, Army officials executing AIA for programs funded and directed by other DoD, or non-DoD, agencies often have little insight into the goals or ultimate results of their efforts. Army implementers often keep track of inputs and outputs, but not outcomes. In such cases, it probably makes most sense for AIA officials to provide data to the organization primarily responsible for the security cooperation program rather than to HQDA. However, that organization may not have a database that is compatible with AIAKSS, and there is currently no guarantee that HQDA would have access to the information being provided by its subordinates to other agencies.

Who Should Provide AIA Program Information?

One of the most difficult problems HQDA will likely face in AIAKSS implementation is the question of who will be responsible for submitting data to the system. HQDA wants to collect program-level data for AIA, but there is no standard definition for an Army international "program" or "activity." Furthermore, security cooperation activities are organized differently across Army organizations. Some have a single "manager" who maintains broad visibility of all program activities. Other organizations lack a well-defined reporting structure. In such cases, AIA officials may have only a general idea of the activities being carried out within their commands. Since AIA programs differ substantially in how they are organized and how they manage their international activities, it would seem appropriate that AIA officials at the MACOM level take responsibility for developing a reporting structure for the programs and activities within their domains. In the process, they will gain a better appreciation for the sources of AIA data available to them as well as the personnel best positioned to collect these data for AIAKSS.

How Should Assessment Data Be Aggregated?

Assessing effectiveness requires careful analysis of how well the results of AIA programs contribute to the Army's goals. With respect to short-term, quantitative outputs, the aggregation process should be rather straightforward. Making use of the "dual-level" collection tool in AIAKSS, program officials would sum the mostly numerical data provided by their subordinate activities in the "field basket" of AIAKSS and enter the information in the system's "program basket." However, in most cases, the process of assessing the long-term, qualitative outcomes of international activities will require more expert judgment on the part of the program manager. In addition, it may be difficult to associate AIA outputs with outcomes in one particular year because it often takes an extended period of time for security cooperation activities to have their intended effect or become evident. For this reason, AIAKSS is built to allow programs to indicate whether currently observed outcomes are derived from activities conducted in previous years. Similarly, a program's reporting of output data for the current year may help to explain outcomes in future years.

How Should Measurement Targets Be Set?

The current configuration of AIAKSS contains only indicators of AIA performance and effectiveness. At least in the near term, there will be no specific measurement targets for which program managers must account—i.e., no MOEs at this stage. This was done deliberately to develop a baseline for measuring AIA. Measurement targets, which prod AIA officials in the direction determined by Army and DoD leadership, will be set in the future against such a baseline. As the Army's authority on international activities, HQDA G-3 must approve any high-level targets that are established for AIA. However, as Chapter Four makes clear, metrics development and the setting of specific goals or targets are best achieved through a consultative process involving both supervising authorities and executing agencies. Such a process helps ensure that targets are in line with organizational objectives and are challenging, while still being feasible and fair, thus reducing the chances that implementers might attempt to "game" the system. Furthermore, HQDA is not the only arbiter of security cooperation activities involving Army agencies. OSD, the Department of State, and the regional Combatant Commands, among others, have important stakes in many AIA and will want to influence the outcome targets for these activities. Also, it is unclear whether HQDA needs to play a role in determining specific targets for immediate AIA performance results (i.e., outputs or MOPs). These are probably better left to AIA program managers or command officials in consultation with their subordinate activities.

Integrating AIAKSS into the DoD Assessment Realm

As the previous section implies, the Army will find it hard to assess its international activities on its own. Too many other agencies are involved in AIA funding and oversight to ignore their requirements and feedback. In addition, the Office of the Secretary of Defense for Policy is seeking a mechanism for integrating the various data collection and assessment mechanisms currently being used, or under development, by the regional Combatant Commands, the services, and other defense agencies.

One potential obstacle to integration is that AIAKSS, with its Title 10 service orientation, supports program assessment, whereas the regional Combatant Commands have been developing country-focused assessment systems. To surmount this hurdle, the Army might consider linking program assessments to specific countries. Such an approach would prove easier for some AIA programs than others. For example, the National Guard's State Partnership Program links states with specific foreign countries, and data are collected by country. But for programs that have a global or cross-regional focus in their activities, collecting country-level data will prove to be more challenging. Adding a country dimension to Army assessment would also require an expansion of the capabilities of AIAKSS.

Figure 5.2 shows the current configuration of AIAKSS in the unshaded areas and possible expanded capabilities in the shaded areas. Country-level assessment is currently limited to identifying where programs operate and how programs rank their objectives by country. Funding is currently tied to the program and not to the country because the Army funds programs and not countries. Currently, AIAKSS asks for a broad assessment that includes general outputs and outcomes for a program. Some country data will naturally be gathered when successes in particular countries are used as evidence for program success. However, currently, outputs and outcomes are not required by country. As Figure 5.2 indicates, it would be possible for HQDA to require that programs provide assessment and funding data by country via AIAKSS. However, it may be that the cost and difficulty of collecting these data outweigh the usefulness of trying to expand the capabilities of AIAKSS.

Another approach would be for the services and the COCOMs to develop separate, but linked, AIA assessment mechanisms (see Figure 5.3). The COCOMs could continue to perform country-level assessments, whereas the services (and possibly, DoD agencies such as DSCA) could develop programmatic assessment tools. For comparative purposes, they could be linked through a common set of high-

Figure 5.2
Possible Expansion of AIAKSS Capabilities

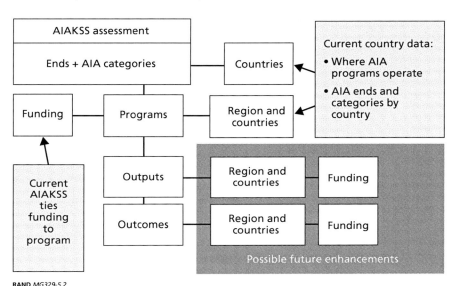

RAND *MG329-5.2*

level security cooperation ends and ways, ideally sanctioned by OSD and the Department of State (DoS) (so as to cover security assistance activities). At the other end of the spectrum, all security cooperation agencies could tap into an integrated set of "event tracker" databases, such as PACOM's TSCMIS or the National Guard State Partnership Program's database.

Conclusion

Whatever the outcome of OSD's efforts to integrate the assessment systems of the services and the COCOMs, it is clear that the security cooperation community is paying increased attention to high-level assessment. No longer is the focus solely on the collection of data on estimated costs and the number of activities or events conducted. There is a growing emphasis on how well activities—whether

Figure 5.3
Separate But Linked Security Cooperation Assessment Scheme

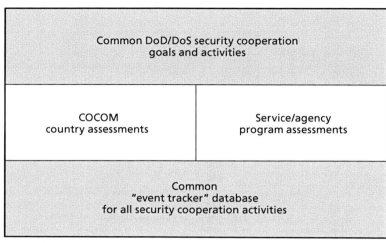

aggregated by program or country—contribute to achieving COCOM, service, and DoD objectives. Our assessment methodology, which has been incorporated into AIAKSS, provides a way for the Army to systematically assess the contribution of AIA to its goals. Yet it is clear that AIAKSS is a tool rather than a magical black box producing answers for every question about the performance and effectiveness of international activities. How well AIAKSS will work to support assessment will depend on the quality of data gathered and the quality of judgments made at several levels of AIA officials. Thus, it will be important to provide training and assistance to system users, to keep the tool updated, and to exercise good management over AIAKSS and the assessment process.

CHAPTER SIX

AIA Test Cases

Introduction

Initial feedback from AIA personnel gathered in the first year of our assessment research in 2003 indicated broad understanding of, and support for, our general approach. However, there were also important questions raised about its implementation, several of which were mentioned in the previous chapter. This chapter presents three test cases that were used to explore the utility and feasibility of our AIA assessment method and collection/reporting tool. The test cases were the U.S. Army Medical Department (AMEDD), the U.S. National Guard Bureau's State Partnership Program, and U.S. Army South (USARSO). Although these cases did not represent a comprehensive validation of our approach, they offered useful insights to support the Army's implementation of AIAKSS.

Two basic criteria guided our selection of test cases.[1] Primarily, we wanted cases that varied in their institutional structure to help us understand the broad spectrum of organizations that implement AIA programs and provide us insights into how different organizations report AIA data. Secondarily, we sought test cases that would capture differences in regional and functional perspectives. AMEDD, SPP, and USARSO appeared to meet these criteria. First, they share few

---

[1] We endeavored to follow the sound selection advice found in Van Evera (1997); King, Keohane, and Verba (1994); and George and McKeown (1985, pp. 43–68).

similarities in how they conduct AIA and how they organize themselves to do so. Second, USARSO is a regional Component Command that focuses its security cooperation efforts in Latin America, whereas AMEDD and SPP are Functional Commands whose medical and state partnering activities are conducted in many countries and regions around the world.

The following three sections describe the international activities carried out by AMEDD, SPP, and USARSO and present the commands' responses to our assessment method and data collection/reporting tool.

U.S. Army Medical Department

The U.S. Army Medical Department is a major Army command that provides a broad range of medical services and related activities to the U.S. Army. It is responsible for the Army's fixed hospitals and dental facilities, its preventive health, medical research, development and training institutions, and a veterinary command that provides food inspection and animal care services for the entire Department of Defense. AMEDD provides trained medical specialists to the Army's combat medical units that are assigned to Combatant Commanders. Most AMEDD deployments are in support of humanitarian assistance, peacekeeping, and stability operations rather than combat operations. Many of these deployments involve Army Reserve and Army National Guard units. In fact, about 63 percent of the Army's medical forces are in the Reserve component.

The Commanding General of the Medical Command (MEDCOM)/Surgeon General directs AMEDD, with support from executive agencies, headquarters directorates, major subordinate commands, and the Surgeon General staff (see Figure 6.1).

AMEDD and Army International Activities

AMEDD is active in many Army International Activities, some of which it funds and implements, and others which it manages for

Figure 6.1
AMEDD Organization

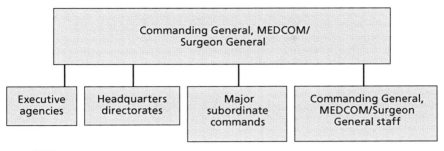

RAND *MG329-6.1*

other Army and DoD organizations. Table 6.1 presents examples and descriptions of AMEDD international activities by AIA way.

Organization of AIA Within AMEDD

International activities mostly fall within the purview of AMEDD's major subcommands. The latter provide facilities and experts to support Army international medical education and training, and research and development activities. They also oversee the transfer of military medical materiel and technical training, participate in international forums on military medicine, and facilitate military exchanges and senior leadership contacts on Army medical-related matters. The regional Medical Commands typically serve as host to international activities, whereas the functional subcommands, such as the Army Medical Research and Materiel Command, are generally responsible for organizing AMEDD international activities. The Office of the Surgeon General (OTSG) is also involved in managing some international activities. These include education and training activities and military contact activities between senior Army medical leaders and their foreign counterparts. In short, Army International Activities cut across the major subcommands (see Figure 6.2).

When asked to consider the applicability of AIA ways in characterizing AMEDD international activities, AMEDD personnel

Table 6.1
AMEDD and Army International Activities

AIA Ways	Examples	Purpose
Education and training	IMET and other programs that enroll "students" in courses at AMEDD schools and AMEDD medical centers	Professional knowledge and social networks that result from such activities help the United States to negotiate for access, improve interoperability and other interactions with foreign armies, and encourage stable leadership transition in foreign armies.
International support arrangements and treaty compliance	Humanitarian Assistance Program	This program fulfills the U.S. Army commitment to provide medical humanitarian civic assistance to populations in foreign countries.
Materiel transfer and technical training	FMS and IMET	Transfers and training improve the quality of equipment and expertise available in countries that are critical to U.S. Army missions and objectives; for AMEDD, these activities are expected to facilitate reduction in excess materiel, improve interoperability, and improve the ability of recipient countries to provide medical assistance to support U.S. Army action in times of need.
Military contacts	Personnel Exchange Program	Senior leadership contacts improve interoperability by facilitating dialogues in setting standards for military and medical technologies, as well as contribute to U.S. Army transformation.
Military exchanges	Army Personnel Exchange Program	Exchanges expose U.S. medical personnel to medical research and care facilities in foreign nations (and vice versa) as well as increase knowledge and capacity, enhance interoperability, and expand institutional contacts for the U.S. Army.

Table 6.1 (continued)

AIA Ways	Examples	Purpose
Military exercises	MEDFLAG, a joint (multi-service) and combined (multicountry) medical training and civic assistance exercise conducted annually by EUCOM	This exercise provides medical training, e.g., medical evacuations, in support of military exercises. AMEDD also obtains medical information and techniques from host nation personnel and builds capacity in telemedicine and mobility for Army medical units.
RDT&E	Engineers and Scientists Exchange Program, Data Exchange Agreements, and Cooperative R&D Agreements with allies and other nations	RDT&E ensures military and medical modernization and facilitates transformation of AMEDD and the U.S. Army. Target countries might adopt best practices to manage their RDT&E enterprises, such as civilian oversight, thus, promoting democracy and stability in the long term.
Standing forums	AMEDD participation in NATO Medical Working Groups (general medicine and nuclear, biological and chemical), Security Assistance Training Program Reviews (of CENTCOM, EUCOM, PACOM), Asian-Pacific Medical Conference, the U.S.-South African Bilateral Defense Committee, and the Hungarian Medical Conference	Participation in these forums is expected to help the Army to acquire/maintain access in foreign territories, encourage standardization agreements for improved interoperability, and enable the U.S. Army to solicit diplomatic support for U.S. policies.

reported that the typology largely coincided with the organization of international activities within AMEDD. The AIA ways and the AMEDD offices responsible for their management are listed below:

- Education and training—OTSG
- International agreement support and treaty compliance—OTSG
- Military contacts—various AMEDD major subcommands
- Military exchanges—OTSG
- Military exercises—Current Operations Office

Figure 6.2
Organization of AIA Within AMEDD

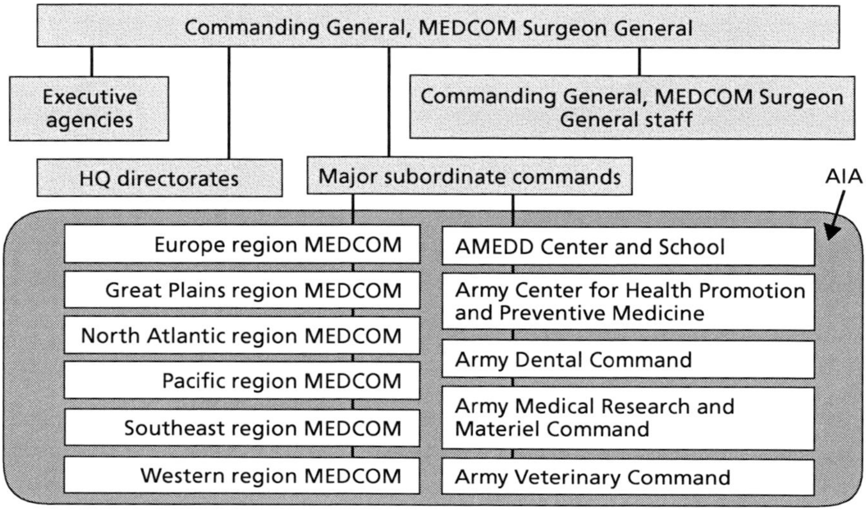

RAND *MG329-6.2*

- Materiel transfer and technical training—U.S. Army Medical Materiel Agency (USAMMA)
- RDT&E—U.S. Medical Research and Materiel Command (MRMC)
- Standing forums—various AMEDD major subcommands.

AMEDD International Activities and AIA Ends

Reflecting their multidimensional character, AMEDD international activities were seen by AMEDD officials as supporting multiple AIA ends and aligning with multiple AIA ways. The following subsections record the responses gathered about several AMEDD international activities within the context of our AIA ways.

Education and Training. Army education and training activities are coordinated by officials in the OTSG and executed at the major subcommand level. One prominent example is the Medical Strategic Leadership Program (MSLP). MSLP is a multiservice and multicountry postgraduate medical program for senior military medical leaders and civilian equivalents. Although the program involves U.S. and foreign personnel from every regional Combatant Command, all MSLP activities are conducted in the Northern Command (NORTHCOM) region. Funding for MSLP students is provided through FMS or IMET.

MSLP has three objectives. The first is to turn participants into strategic health care executives in their own countries. The second is to help participants improve their ability to work within coalitions and alliances. The third is to expand the understanding of, and concern for, human rights among participants.

Given these objectives, AMEDD personnel placed MSLP within the AIA categories of "education and training" and "military contacts." In their view, MSLP supported the AIA ends of improving defense capabilities, interoperability, and cooperation as well as helping the Army to establish relations with foreign military forces.

MSLP sessions are conducted three times a year and each lasts three weeks. Each iteration focuses on a particular region of the world. Typically, the first week involves only foreign participants. They are introduced to AMEDD and sit in on briefings and discussions about finance, personnel, logistics, and telemedicine. This segment is tailored to the specialty and specific country needs of the participants. Participants also spend a day at Fort Hood, Texas, to experience life on a large Army installation, meet with the Corps Surgeon, and observe field medical equipment.

Senior U.S. officers from all AMEDD branches and the U.S. Army join the class in the second week. Major areas of instruction include strategic planning and decisionmaking, military medical readiness, leadership in coalition health-service support operations, task force management, international law, state and federal agency interaction, disaster preparedness planning, and interaction with nongovernmental agencies. Guest speakers include regional Combatant Commanders and Joint Task Force surgeons, as well as participants in United Nations Peacekeeping Operations and Operations Other Than War. Participants learn about National Disaster Medical Systems management and various types of multinational contingency planning support operations, including humanitarian assistance. International students provide an informal presentation on their military organization and their leadership challenges.

The third week of the program takes place in Washington, D.C., where foreign and U.S. participants visit DoD health care facilities and simulation centers. They meet with the Army Surgeon General and the Joint Medical Staff. Additional visits are scheduled to area facilities, including the Center for Health Promotion and Preventive Medicine, the Walter Reed Army Institute for Research, the U.S. Army Medical Research and Materiel Command, and the Office of Homeland Defense. In addition, visits with members of Congress, a call to the U.S. Department of State, and a meeting with members of the national media represent the political dimensions of this program.

Materiel Transfer and Technical Training. Materiel transfer and technical training activities are conducted under the auspices of USAMMA, the Army Surgeon General's executive agent for strategic medical logistics programs and initiatives. USAMMA's mission is to enhance medical logistics throughout the full range of military health to service support missions worldwide, develop and implement innovative concepts and technologies, and advance medical logistics information and knowledge.

USAMMA officials indicated that their programs were associated with several AIA ways and supported multiple AIA ends. The FMS program, for example, was perceived as helping the Army to establish relationships, promote cooperation, and enhance interoperability. Relevant ways and ends are shown in bold in Table 6.2.

USAMMA works closely with the DSCA under the Office of the Secretary of Defense. DSCA is the DoD office with primary responsibility for the FMS program and other security assistance programs. FMS covers all government-to-government purchases of weapons and other defense articles, defense services, and military training. All foreign purchase requests must go through DSCA. In the medical area, DCSA procures materiel and technical training from AMEDD through USAMMA. The latter then sends the medical materiel to foreign buyers or receives foreign personnel for medical logistics training.

Table 6.2
FMS Program—Relevant AIA Ways and AIA Ends

AIA Ways	AIA Ends
Education and training	Ensure access
International support arrangements and treaty compliance	Promote transformation
Materiel transfer and technical training	**Improve interoperability**
Military contacts	Improve defense capabilities
Military exchanges	Promote stability and democracy
Military exercises	Assure allies
RDT&E	**Improve non-military cooperation**
Standing forums	Establish relations

Although FMS is the main source of funds for medical materiel transfers and technical training, USAMMA also infrequently provides such training for foreign personnel under the IMET program. Requests for this type of technical training are typically routed to USAMMA through the OTSG. Participants are assigned to the appropriate offices within USAMMA or major subcommands within AMEDD to receive technical training. USAMMA personnel indicated that their agency's activities were linked with several AIA ways and AIA ends. These are shown in bold in Table 6.3.

Research, Development, Technology, and Engineering. RDT&E activities are managed by MRMC within several programs: the Engineers and Scientists Exchange Program, the Personnel Exchange Program, and National Research Council fellowships. Like other AMEDD international activities, RDT&E activities were seen to support multiple objectives. For example, according to MRMC officials, the Personnel Exchange Program facilitates higher-level interaction and supports Army medical RDT&E through access to foreign medical resources, including research facilities and data, expertise, technologies, and cooperation in conducting vaccine trials.

Table 6.3
IMET—Relevant AIA Ways and AIA Ends

AIA Ways	AIA Ends
Education and training	Ensure access
International support arrangements and treaty compliance	**Promote transformation**
Materiel transfer and technical training	Improve interoperability
Military contacts	Improve defense capabilities
Military exchanges	Promote stability and democracy
Military exercises	Assure allies
RDT&E	**Improve non-military cooperation**
Standing forums	**Establish relations**

The decision to enter into RDT&E collaboration with foreign militaries typically lies with MRMC research area directors, who are directly involved in the day-to-day management of RDT&E activities in research laboratories. Regular meetings between MRMC program officials and research area directors facilitate communication and coordination in supporting visits by foreign researchers and ensuring their placement in appropriate locations within the regional MEDCOM and the functional subcommands. In most instances, the MRMC program funds foreign medical RDT&E visits and collaboration. However, research area directors also control funds for international activities that come from 6.1 and 6.2[2] research funds, the Defense Health Care Program, and other sources.

Military Contacts and Standing Forums. AMEDD international activities, related to military contacts and standing forums, include visits and meetings held under the auspices of NATO; American, British, Canadian, and Australian (ABCA) Armies' Program; and the

---

[2] DoD uses these S&T activities for descriptive and budgeting purposes. Both 6.1 and 6.2 include costs of laboratory personnel, either in-house or contractor-operated. Basic research or 6.1 supports systematic study directed toward greater knowledge or understanding of the fundamental aspects of phenomena or observable facts without specific applications toward processes or products in mind. Applied research or 6.2 supports systematic study to gain knowledge or understanding necessary to determine how a recognized and specific need may be met.

U.S.-South Africa Bilateral Committee. Activities in these two categories are particularly difficult to track because of their large numbers and the many AMEDD offices involved in their planning and execution. Ranging from high-level meetings between the Army Surgeon General and his foreign counterparts to professional conferences in the United States and overseas attended by individual AMEDD scientists, these activities vary considerably in their visibility and importance. In some cases, AMEDD officials do not even record participation in professional conferences in their tally of international activities. Also, funding for them derives from a variety of sources and is channeled through many offices.

Metrics and Indicators

In general, AMEDD personnel thought that the AIA ways, ends, and indicators in our assessment methodology were useful in conveying the products and accomplishments of their programs. For example, MRMC officials said that counting the number of scientists and engineers participating in RDT&E activities was a good way to express the extent of the relationships built through these activities. The development of scientific and technical relationships might, in turn, enable international collaborative efforts and improve U.S. access to foreign research data and technologies, both of which contribute to the Army's ultimate security cooperation ends.

However, AMEDD officials expressed concern that evidence for AIA indicators might not be easily available to personnel assigned responsibility for submitting data to AIAKSS. Managers of international activities at AMEDD do not necessarily understand their rationale, their immediate results, or their ultimate effect on the Army's security cooperation ends. In many cases, AIA officials would need to solicit input from their colleagues within or outside AMEDD to gather this information—a time-consuming task, particularly when the policies and procedures are not currently available to support such interoffice queries. Further complications could arise if the appropriate office or agency refused to share its data or could not provide the requested information because they did not maintain it. Over time, new processes and relationships could be developed to support comprehensive AIA assessment at AMEDD. Until then, however, the onus for providing such assessments will fall on AMEDD personnel with limited time, understanding, and resources.

Funding Information

Tracking the resources for AMEDD international activities will be difficult because those responsible for AIA planning and execution do not always have control, or even visibility, over funding. In RDT&E, for example, MRMC program officials reported that although they are responsible for the day-to-day management of RDTE activities, funding decisions and data belong to research area directors and their programming office. The same is true for USAMMA officials involved in medical materiel transfer and technical training activities. Another challenge to obtaining funding data is that activities hosted by AMEDD are not funded by AMEDD. AMEDD receives reimbursements from other Army or DoD organizations for hosting U.S. and foreign participants in AMEDD international activities. AMEDD is not generally aware of the funding source. Moreover, even if AMEDD had this information and could submit it to AIAKSS, there could be a problem with double-counting should AMEDD and its sponsor both report on the same activities.

AMEDD faces several other difficulties in collecting AIA resource information. The funding data typically available to AMEDD managers indicate how much an activity costs, rather than the actual dollars spent. Also, personnel time constitutes the main resource input for many international activities, e.g., international conferences, meetings, and visits. And except for international program officials, the personnel costs of AMEDD's international activities could not be accurately determined. Finally, associating AIA funding with security cooperation ends is difficult when Congress directs such funding for particular activities, not Army objectives.

Key Insights

The AMEDD test case provides some useful insights for our AIA assessment effort. First, as a major Functional Command, AMEDD "owns"—that is, it funds, plans, and executes—only a small handful of AIA programs, such as MSLP. More typically, it supports international activities that are owned by other Army or DoD organizations by providing facilities, materiel, expertise, contacts, and other AMEDD assets and is reimbursed for what it provides. This means that the motivations for these non-AMEDD-owned international activities and their funding, and information about their results, are not always visible to AMEDD personnel, which hinders their ability to submit data to AIAKSS.

Second, it appears reasonable to request a MACOM, such as AMEDD, to provide AIA information on programs it does not own, but this could create a problem if more than one organization were to submit data to HQDA on the same activities. One way to avoid duplication would be to ask Army organizations to only assess, and directly report on, programs they own. For programs that they support by providing facilities, expertise, etc., they would provide data to the program owners to use in their assessment and submission to AIAKSS. However, such a requirement has its own weakness: It would mean that the Army's contribution to programs that are not owned by any Army organization but which the Army supports (e.g., IMET and FMS) would not be fully apparent.

Third, AMEDD does not currently have a central repository of information on international activities. In general, AIA officials who manage AIA coordinate only with those organizations necessary to accomplish their work. Moreover, requirements for internal reporting or information-sharing—when they exist—are not always practiced or enforced. A first step, therefore, might be to encourage offices to comply with existing regulations by building mechanisms and processes to support them when necessary and providing incentives to do so where possible. In the case of AMEDD, there are clear lead offices for some types of international activities: e.g., MRMC for Army medical RDT&E and USAMMA for medical materiel transfer and technical training. Responsibility is more dispersed among AMEDD offices for other categories of activities, such as military contacts, military exchanges, and standing forums. In these cases, AMEDD may need to designate a single office as the repository for AIA information to facilitate data collection and reporting for AIAKSS.

National Guard Bureau State Partnership Program

The National Guard Bureau (NGB) State Partnership Program (SPP) has been operating since 1993. Originally employed as a way to establish post–Cold War links between NGB personnel and East European militaries, the program has now expanded into Asia and Latin America.[3] Today, 39 U.S. states, two territories, and the District of Columbia are partnered with 44 countries around the world. The state partners are shown in Figure 6.3.

The SPP conducts a wide range of activities that cover every category of AIA. These activities are listed below:

- Professional military education
- Command and control command post exercises

Figure 6.3
State Partnership Program

SOURCE: This map is updated frequently and can be found at http://www.ngb-ia.org/public/spd.cfm/spi/states_map.
RAND MG329-6.3

---

[3] For a detailed study of the role of the National Guard Bureau State Partnership Program in implementing U.S. security cooperation activities, see Howard (2004, pp. 179–196).

- Small unit exchanges
- Consequence management
- Environmental management and education
- Military medical exchanges
- Civilian medical exchanges
- Civilian education exchanges
- Public affairs
- Search and rescue (command post exercise and actual)
- Emergency preparedness
- Counter drugs
- Humanitarian construction
- Border control
- Senior leader development
- Civic leader development.

Organizationally, the SPP is well integrated into the COCOM planning structure. For example, Guard personnel fill the EUCOM new "Bilateral Affairs Officer" position on each embassy team, and they are responsible for facilitating non-combat security cooperation in Eastern Europe and Central Asia. In addition, the SPP has established a full-time coordinator for each state. This hierarchical structure makes the overall SPP more coherent than most security cooperation programs. Finally, the Guard seamlessly incorporates both Army and Air National Guard forces.

Current Assessment
Currently, the SPP assesses its overall effectiveness mainly through anecdotal evidence. However, the NGB is moving in the direction of systematic assessment with a newly deployed "event tracker" database, which monitors each security cooperation event from its conception until its execution. This is no small feat, since four different organizations must approve of an event before it is funded and executed: the state, the National Guard Bureau Headquarters, the U.S. embassy in the recipient country, and the headquarters of the appropriate U.S. Combatant Command. The event tracker assesses events by means of After-Action Reports (AARs). Although they provide some of the

most powerful arguments for the conduct of security cooperation activities with foreign militaries, these AARs have not yet been organized into an overall program assessment.

Funding

The SPP has a very complex funding picture. Congress directly allocates the National Guard Bureau approximately $2 million per year to help fund some SPP initiatives. All other funding comes from the states, the Combatant Commands' traditional commander's activities (TCA) accounts, the Department of Defense, and the Department of State. The SPP has deliberately chosen to fund activities in this manner. This keeps the ownership and planning for SPP activities largely in the hands of the COCOMs and ensures that all activities are planned with COCOM theater security cooperation ends in mind. The downside to this funding strategy is that the NGB has minimal control of the future of the SPP. For now, the Guard is satisfied with this arrangement, but it does cause some problems. For example, there is some confusion over who should assess the effectiveness of SPP activities. Since the COCOMs are mostly responsible for SPP planning and funding, some argue that assessment should be left to them. However, National Guard Bureau officials must defend SPP activities to their leadership and to Congress. So there is a need for the Guard to evaluate its effectiveness at implementing the ends that other organizations have established for SPP.

AIAKSS Test

While the electronic version of AIAKSS was still under development, a paper version of the database and reporting tool was presented to selected SPP officials during a EUCOM-sponsored conference in Stuttgart, Germany, in March 2004 for SPP state coordinators and bilateral affairs officers. Over the course of three days, eight state coordinators were asked to identify where their programs operated, choose which of the eight AIA ends were pertinent for their country, and then rank the ends in order of importance. Because they were new in their jobs, some respondents did not feel comfortable assessing high-level security cooperation ends. Others did not think it was the responsibility of the state coordinator to conduct assessments. How-ever, after encouraging coordinators to confer with their Bilateral Af-fairs Officers, most state coordinators were able to rank the ends for their countries.

Not surprisingly, the majority of respondents chose "establish relations" as their most important AIA end (see Figure 6.4). This makes sense, since the SPP is used by EUCOM as a way to establish relations with new countries that have not had ties with the U.S. mili-tary in the past. This is especially true for Central Asian and African countries that are sensitive to being seen as cooperating with the U.S. military. Because of the SPP's ability to focus on non-combat activi-ties, such as disaster assistance training and development of non-commissioned officers, those countries tend to view Guard activities as helpful and non-exploitative.

Table 6.4 shows that at least two of the respondents chose all of the ends. This reflects the fact that some of the more established partnerships have moved beyond the "establish relations" end and into more advanced ends such as "improve interoperability."

Figure 6.4
Number of SPP Partner States That Selected Each AIA End

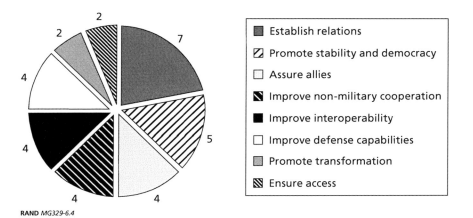

RAND MG329-6.4

Table 6.4
Number of States That Selected Each AIA End

Establish relations	7
Promote stability and democracy	5
Assure allies	4
Improve non-military cooperation	4
Improve interoperability	4
Improve defense capabilities	4
Promote transformation	2
Ensure access	2

The respondents also identified four relevant AIA ways. These were military-to-military contacts, military-to-military exchanges, military exercises, and education and training. The respondents were asked to link the type of activities conducted to the ends they had chosen earlier. State coordinators identified military-to-military contacts as the most important way to establish relations with a foreign country. Military-to-military exchanges, as well as military-to-military contacts, were linked to improving interoperability. All of the selections of the state coordinators with respect to AIA ways and ends are shown in Table 6.5.

Without the assistance of an electronic database, the process of linking output and outcome indicators to AIA ways and ends was rather grueling. In the end, we were able to collect information from only three states. Figure 6.5 shows the output and outcome indictors selected by these three states for their most important AIA end (establish relations) and most important AIA way (military-to-military contacts).

AIAKSS prompts respondents to choose one or more indicators for each end and way combination. In this case, two output and two outcome indicators were chosen for the end of establish relations. The number of contacts with a target country was associated with establishing institutional points of contact and preparing a nation for peacekeeping operations. The "highest rank" indicator was developed to show how high-level contact activities differ from regular contact

Table 6.5
SPP Selections of Ways and Ends

	AIA Ways			
	Military-to-Military Contacts	Military-to-Military Exchanges	Military Exercises	Education and Training
Establish relations	6	1	1	1
Promote stability and democracy	4	1	1	
Assure allies	3	1		
Improve non-military cooperation	3	1		
Improve interoperability	4	3	1	
Improve defense capabilitie	2	1	1	
Promote transformation	2	1	1	
Ensure access	2		1	1

Figure 6.5
Chosen Output and Outcome Indicators for Establishing Relations and Military Contacts

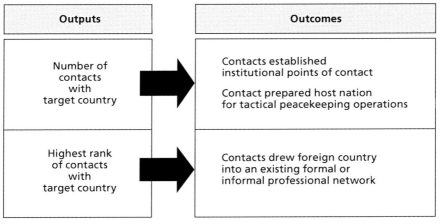

activities. In this case, a high-ranking U.S. officer was tapped to help bring a new country into NATO. High rank individuals can be used to start off a new relationship or to help mark significant milestones in the development of a bilateral relationship. Assessments of the utility of such contacts could be important if policymakers are trying to decide what level of representation is needed in a particular country, or if costs are a concern. High-ranking visits are expensive, and a lower-ranking delegation can be equally or more effective, depending on the purpose of the trip.

Key Insights

The NGB test case provided us with some additional issues to consider when implementing AIAKSS. We learned, for example, that

- Most state coordinators do not have access to SIPRNET. This led us to recommend to HQDA that AIAKSS be placed on the unclassified NIPRNET until most AIA officials are linked to the military's classified Intranet.
- Even though each state has a designated SPP coordinator, several of our respondents did not feel comfortable providing the requested AIA information to AIAKSS. New coordinators did not know which ends their program supported, and they were unable to choose indicators without help from their associates in the embassies. After the coordinators were encouraged to contact their Bilateral Affairs Officers, who are National Guard personnel assigned to certain countries in Europe, they were able to select appropriate ends and indicators.
- Despite their difficulties in selecting indicators, SPP coordinators preferred to choose indicators from a predetermined list.
- State coordinators thought that the AIAKSS tool might help them to report the results of their efforts to their respective governors and state Guard officials.

U.S. Army South

As the Army Component Command for SOUTHCOM, USARSO's area of responsibility is Latin America.[4] The G-5 Civil Affairs section of USARSO conducts the bulk of these activities. To formulate their country plans for security cooperation, they rely on guidance primarily from SOUTHCOM but take guidance from HQDA as well. To support the Combatant Commander, USARSO staff members coordinate their plans closely with their counterparts in the J-5 at SOUTHCOM. The bulk of USARSO's international activities involve regularly scheduled exercises, exchanges, and standing forums. Most of the funding for USARSO's involvement with Latin American militaries comes from SOUTHCOM's TCA account.

The J-5 directorate of SOUTHCOM conducts most of the assessments of the security cooperation activities in Latin America. However, the most sophisticated of these assessments focus on the evolving military capabilities of countries of emphasis.[5] At the same time, there is an increasing recognition by USARSO that assessing security cooperation could improve its planning process. For example, members of the G-5 staff have begun to develop measures of effectiveness to evaluate their annual exercises. Their hope is to find a method that will help them use annual exercises to improve the capabilities of participating countries over time. Although a process for doing this has not been formally established, there is a growing awareness at USARSO that measures of effectiveness could prove useful in allocating scarce resources for exercises. Although G-5 staff at USARSO devote most of their attention to implementing SOUTHCOM's Theater Security Cooperation Plan, they also rely heavily on two Army programs to support the COCOM's and Army's objectives.

---

[4] With the move from Puerto Rico to San Antonio, Texas, USARSO has also become a major subordinate command under Forces Command (FORSCOM). In this way, USARSO is both subordinate to FORSCOM as well as acting as the Army component for SOUTHCOM.

[5] OSD and the COCOMs have specialized guidance for selected "countries of emphasis" in each region.

Key Insights

The USARSO test case provided a few insights that might affect the implementation of AIKSS. First, in USARSO's view, the "management" of Army programs is not all that clear, and HQDA may be the best place to assess the effectiveness of these programs. Second, Component Commands such as USARSO seem primarily concerned with implementing the Combatant Commander's Theater Security Cooperation Plan. As a result, they take a country-specific view when assessing the effectiveness of security cooperation. In their view, asking about the effectiveness of specific programs is understandable but contrary to their principal mission.

Conclusion

Although AMEDD, SPP, and USARSO differ greatly in the scope and focus of their involvement in Army International Activities, none of the participants in our three test cases objected to our basic assessment approach, and they reported similar challenges in implementing the assessment methodology in AIAKSS.

First, an Army organization that supports an international activity is not always cognizant of the reasons for undertaking it or privy to the funding sources that are supporting it. Neither are they always aware of the results, or more importantly, how those results relate to the Army's security cooperation ends. This is because Army organizations often host international activities and provide necessary expertise or materiel, but they do not always participate in decisions related to security cooperation planning, funding, and assessment. These decisions are made by organizations that "own" a program or activity, not by service providers.

Second, our test case participants experienced great difficulty in determining who should analyze and aggregate AIA data and submit assessments to HQDA. "Program manager" was a term used loosely to describe such a person, but many programs do not have a single program manager. This is because the execution of an AIA program often requires the assistance of many organizations that possess the necessary expertise, materiel, facilities, and other assets to enable the conduct of a program's activities. This suggests that although those who plan or fund a program presumably know its ends, they do not always know its results when program activities are conducted by other organizations. Similarly, those who support the implementation of a program might observe the results but would not necessarily know whether those results align well with the ends of the program planner.

CHAPTER SEVEN

Concluding Observations

Analysis of our three test cases, as well as the feedback we have received from other Army organizations, shows that our basic AIA assessment approach appears sound. In addition, there is a consensus that a web-based database and reporting tool, such as AIAKSS, is needed to capture and share AIA-related programmatic, funding, and assessment information throughout the Army and DoD. Nevertheless, there is also recognition that the Army, in cooperation with the rest of the security cooperation community, must overcome some major hurdles before it can conduct a comprehensive and objective evaluation of its international activities.

In this chapter, we summarize the lessons we have learned from developing, testing, and refining our AIA assessment approach with the assistance of AIA officials in HQDA and throughout the Army. In addition, we discuss ways the Army might use the information collected by AIAKSS for analytical and assessment purposes. We close the chapter by highlighting several important issues that need to be considered before the Army can fully and effectively implement our assessment model.

AIA Lessons Learned

AIAs Are Highly Diverse

The Army's international activities are highly diverse not only in the variety of programs involved but also in the broad array of activities that take place in each program. For example, Army international medical activities span every category of activity, from education and training activities, to exercises, to high-level military contacts. In our initial effort to categorize AIAs into eight groups, such diversity presented a challenge. It also forced us to think hard about how performance measurement should be implemented. The major questions were whether security cooperation programs should be included in more than one category and who should make that decision?

Regarding the first question, we decided that programs could be included in more than one category. The possibility of multiple listings acknowledges the sheer diversity of AIA programs. It also enables us to elicit data that can help show the multitude of ways diverse programs can contribute to AIA ends. On the second question, we decided that it is best for programs themselves to decide which categories they belong to, as well as how many. Individual programs know best what they do and the ways they contribute to AIA ends. Programs should also be allowed to switch categories to reflect changes in their activities. We are inclined to let programs make this decision also because that makes them responsible for providing the evidentiary data needed to back their claims.

Connecting Individual AIAs to Specific AIA Ends Is Not Easy

Army International Activities can be difficult to link to overall security cooperation objectives. For example, how does professional education and training directly contribute to the AIA end of access for the U.S. Army? How can any activity, for that matter, claim to have encouraged democracy and stability in a partner country? Whether new skills and knowledge acquired by foreign personnel will create outcomes that further any of the AIA ends will depend on a host of factors such as placement and promotion, which are affected by political considerations, personality matches, organizational needs, and resource availability. The same is true for scientific discoveries and technological advances, materiel transfer and training, and every other AIA category. Practically every activity and its outputs have the potential to create outcomes that contribute to the AIA ends, but using those outputs to create useful outcomes demands more than what programs and participants can do on their own.

There are other problems with connecting AIA ways and ends. Most Army activities are very specific, whereas Army and national goals are broadly defined. For example, although professional education and training programs build human capital to support each AIA end, it is harder to show the extent to which a particular training activity prepared an individual foreign officer for a certain critical task that affected a specific U.S. Army objective. Other factors beyond the control of the AIA (e.g., a foreign Army staff's placement and promotion policies, as well as political considerations) can also significantly influence how the activity and its outputs are transformed into desired outcomes that will support the Army's AIA ends. A British study that tried to mathematically determine the effectiveness of individual activities in achieving larger national goals failed because the external factors involved overwhelmed the significance of the individual activity.[1]

We decided to approach this problem of causality by selecting indicators that furnish, or help us elicit, information about activities and their outputs. Evidence provided by these indicators cannot tell the whole story, but it can lead to other information that is potentially available for further exploitation. This, in turn, can lead to the establishment of outcomes that contribute to the AIA ends we postulated. We also decided that these ends should provide a reasonable sense of direction or expectation for programs. It is up to the programs themselves to take the next step and begin changing their cur-

---

[1] Email discussion with Dr. Andrew Corcoran on British Ministry of Defence efforts to measure effectiveness of military international activities, March 28, 2003.

rent reporting systems, which are based on compiling outputs, into assessment systems based on specified plans to achieve desired outcomes. Outcomes are defined in terms of the contribution particular activities can make to the various AIA ends.

In sum, our approach centered on finding a way to gauge whether—and, if so, how well—AIAs contribute to AIA ends. The method we developed both identifies and exploits a number of indicators along the path to those ends. The ends themselves play a key role in helping us define and refine the various indicators, particularly outcome indicators.

Combining Quantitative and Qualitative Data Is Necessary

Our literature review and conversations with program officials underscored the value of quantitative and qualitative data in designing a performance measurement system for AIA. We also found that output data are more likely to be quantitative, whereas outcome data are more likely to be qualitative. The challenge lay in making a logical connection between the two kinds of data when dealing with output and outcome indicators. We asked ourselves, for example, if additional (quantitative) units of a certain output necessarily translate into a greater (qualitative) effect or more "useful" outcome when measured against a given AIA end.

What we learned was that in some instances, having additional units of output can improve a desired outcome's support of an AIA end. For instance, having more people trained to know about a certain subject expands the human capital pool, as well as the social networks, available for the Army to draw on in supporting a variety of AIA ends. In similar fashion, having more uniform standardization agreements signed is a first step toward the end of improved interoperability. But transforming outputs into outcomes is not an automatic process, no matter how much output is available. For example, if the Army is in search of cutting edge technology to speed transformation, reaching a technology transfer agreement with a foreign country is the output desired in cooperative science and technology activities. In this and other instances, increasing the number, value, or frequency of activities may lay the foundation for certain desired outcomes, but they do not guarantee them. How security cooperation activities are used proactively and creatively to achieve the ultimate goals is what really matters.

Inputs Are Not the Key to Assessment

"What do we get for the dollars we spent?" is the basic question of interest to the Army and OSD when thinking about international activities. However, an emphasis on tying inputs (dollars and people) to outcomes does not really answer the question of effectiveness, that is, how do international activities advance the Army's security cooperation goals? To begin with, not all inputs can be easily calculated. Inputs in kind, such as personnel support provided in conjunction with other activities, are difficult to estimate. More important, investments in international activities frequently produce payoffs over time. Thus, the Army "buys" immediate results such as foreign military graduates, senior officer visits, or materiel transfers, but the true strategic effect of international activities may become obvious only over time as the training provided, visits exchanged, and materiel transferred create payoffs for the U.S. Army in the form of usable international relationships, capabilities, and access. Therefore, given finite resources, improving the collection and reporting of data related to the outputs and outcomes of international activities, rather than the inputs, may be the best way to support AIA assessment and increase AIA achievements in the long run.

Find Outcomes That Promise to Have an Effect over Time

In our approach to developing a performance measurement system for AIAs, we associate the concept of socialization with outcome indicators. Finding viable outcome indicators that seem likely to have an effect on AIA ends over time is a difficult task. We learned from many conversations with Army personnel involved with AIAs that people choose to work with each other because of a shared experience, goodwill, and trust, and it is people who solve problems, not programs or technologies. Having mechanisms available to help people find points of contact, collaborators, leadership support, and resources is critical to creating effects that will benefit the goals of their organizations and the Army as a whole. Hence, we sought to identify such mechanisms where possible, and they serve as our outcome indictors where appropriate. For example, alumni networks can expand professional links among people involved in AIA. Such networks can help open doors, solve problems, and address issues that otherwise might hinder achievement of the AIA ends.

AIAs Can Have Cross-Cutting Effects

As noted above, many AIAs are diverse in nature, providing a variety of different ways to advance the AIA ends. This diversity also suggests that AIAs can have cross-cutting effects on AIA ends. For example, a multilateral forum enables the U.S. Army to establish new foreign points of contact, which serves AIA ends. The interaction that follows helps the United States and other countries identify and address capability gaps that benefit the AIA end of interoperability. Such foreign contact might also lead to military exchanges that benefit the AIA end of access. Thus, the cross-cutting potential of a single AIA cannot be underestimated, yet it is difficult to envision all the potential pathways this AIA could open, much less claim credit for particular outputs and outcomes. The cross-cutting potential of AIAs also has a temporal dimension. AIA linkages may be exploited over time, contributing to outputs and outcomes that emerge only after years or decades. For this reason, only through concerted efforts to gather AIA data can HQDA appreciate cross-cutting activity linkages and assess how security cooperation benefits the U.S. Army, as well as U.S. allies and partners.

This observation also underscores the importance of information sharing across AIA programs. Giving AIA programs access to information about the outputs and outcomes of all AIAs will help programs use their own information more systematically, coordinate with others, optimize resources, and maximize the effect of outputs and outcomes.

Distinguishing Between Outputs and Outcomes Is Critical

One fundamental challenge to our research team was to distinguish between the outputs and outcomes of AIAs. We decided to designate as outputs those results generated directly by an international activity and outcomes as those that result from the application of the output. This helped us in most instances, but there were some situations in which we questioned whether an output might turn into an outcome, or vice versa, and what would justify a particular classification (or re-classification).

For example, a standardization agreement that is signed as a re-sult of multilateral negotiations is an output. Once this agreement is adopted and implemented, it should be reclassified as an outcome in the next performance measurement reporting cycle to reflect mainte-nance of a level of U.S. commitment to this AIA and its associated AIA end.

By contrast, an outcome indicator such as alumni networks will always remain an outcome indicator in our performance measure-ment system because it can be used again and again to advance the AIA ends, even if networks change in size, shape, or purpose or ex-hibit different levels of effectiveness. For this reason, it should not be removed or reclassified as in the previous example.

Knowing When an End Has Been Achieved Is Important

Some ends are easier to identify as "achieved" than others. AIA ends such as "establish relations" and "ensure access" are fairly easy to gauge. For example, a new foreign point of contact emerges, or the U.S. Army acquires access rights for training or basing in a foreign territory. However, AIA ends such as "assure allies" and "improve non-military cooperation" are more difficult to assess. For example, is a foreign country's trust and confidence in its relationship with the United States increasing? What is a satisfactory level of cooperation for both countries? Knowing when an end has been achieved has im-portant ramifications for deciding how much and what kind of out-puts and outcomes are needed to attain the end.

Feedback from Army personnel and other security cooperation experts indicates that assurance, democracy, and cooperation consti-tute an ongoing *process* as much as they represent AIA ends. For ex-ample, assurance requires a stable and continuous level of resource commitment to, and presence in, international forums. Stability and continuity send signals (shorter-term outputs) and sustain team spirit and enterprise (longer-term outcomes). Also, assurance is a means to achieve other AIA ends. Assuring foreign partners of the U.S. commitment to a relationship with them should help the United States secure support for other AIA ends such as access and interoperability.

Recognizing Failures and Negative Outcomes of AIAs Is Necessary
When data show that an AIA program is unable to convert its outputs into outcomes supportive of AIA ends, what should we conclude? Would it be fair to consider this a failure if it is beyond the program's powers or resource limitations to achieve the desired outcomes? Such a situation is likely in the early years of performance measurement when programs will need time to adjust to the new system and will have to submit data to demonstrate progress toward AIA ends.

Success in implementing performance measurement and building an AIA database suggests that HQDA should assure programs that performance measurement is more about improving how things are done in the future than about evaluating what has happened in the past. HQDA should also work hard to understand what problems might occur in transforming outputs into outcomes. Toward this end, we decided to add a section to the reporting tool that solicits information from programs on potential barriers to their making successful contributions to AIA ends.

An associated challenge is how to account for unintended outcomes of Army security cooperation activities. For example, previous RAND research has shown that training foreign military students in the United States to increase their proficiency in English can lead to an outflow of foreign students from the military into the private sector.[2] As a first step, the AIAP reporting tool we designed asks AIA programs to report "barriers to success" in security cooperation. Should reporting under this heading prove insufficient to account for

---

[2] See Taw (1993, p. 10).

unintended consequences, such as the example just cited, a more pointed question or section could be added to the reporting tool.

Measures Will Need Modification over Time

It has been said that change is the only constant in life. If so, this is also true for a good performance measurement system. Any system that does not evolve would not be able to accommodate changing needs or reflect changing realities. The Army's objectives for security cooperation—i.e., AIA ends—might change over time. In any event, AIA programs are likely to be created, expanded, downsized, or terminated.

The output and outcome indicators we propose for consideration at this time are the result of many hours of research and analysis, as well as valuable input and feedback from Army personnel. They are surely not perfect, but they represent a central component of our answer to HQDA's call for a method to objectively assess the Army's security cooperation activities.

Improvements to these indicators will undoubtedly result from futher consultations among HQDA, AIA officials, and other stakeholders, as well as from implementation of our performance measurement and assessment system.

How Should AIAKSS Information Be Used?

Having taken the aforementioned lessons to heart and incorporated our AIA assessment framework into AIAKSS, there remains the question of how HQDA should use the AIA information that will be reported by AIA officials in the MACOMs. Although this is for G-3 to decide in consultation with the rest of the AIA community, we envision a range of possible uses for these data. For example:

- To meet the metrics requirements of the OSD Security Cooperation Guidance, the Army Strategic Readiness System, and in the future possibly, the OMB's PART;

- To demonstrate to senior Army leaders and national authorities how AIA contributes to service and national security objectives;
- To indicate which AIA operations must be transformed to improve security cooperation outcomes;
- To understand the full range of AIA available to achieve Army and national goals;
- To account for AIA personnel and funding and suggest where scarce AIA resources might be expended to achieve greater payoff;
- To identify and leverage "under the radar" programs in which the Army participates but does not have direct oversight of management or resources;
- To uncover barriers that may be hindering or preventing the successful implementation of AIA programs.

Another opportunity for employing AIAKIS would be as a tool in analyzing the Army's progress in the area of Multinational Force Compatibility (MFC). Developing a strategy for nurturing potential military partners is becoming an increasingly important objective for the Army as coalitions become more ad hoc in nature and the pool of possible coalition partners grows to include many less-capable armies that are not long-time allies. A broadly distributed, high-level data collection and reporting tool such as AIAKSS could be helpful in answering questions regarding

- The appropriate mix of AIA being used to build MFC;
- Major obstacles to building MFC; and
- Relevant measures for evaluating AIA effectiveness in building MFC.

Issues in Applying AIAKSS

From our test case studies and less formal discussions with AIA officials, we discovered several organizational issues that need to be confronted before AIAKSS can be fully employed for analytical and assessment purposes. These issues are listed below. Although we have included some preliminary thoughts on how to deal with them, their resolution will require considerable leadership, flexibility, and coordination on the part of security cooperation officials at the program, command, HQDA, and DoD levels.

First, Army organizations differ greatly in their management of international activities. Functional Commands, such as AMEDD, tend to follow a decentralized approach to the execution of AIA programs, whereas the security cooperation efforts of regional Component Commands, such as USARSO, are tightly coupled to their Combatant Commander's Theater Security Cooperation Plan. As a result, the assessment perspectives of Functional and Component Commands tend to be different, with the former focused on the inputs and outputs of particular global programs and the latter on the outputs and, to a lesser extent, outcomes of numerous, regionally oriented programs. HQDA can either assume the job itself of integrating functional and regional security cooperation perspectives, or it can assign the "ownership" of different AIA programs to specific commands, requiring them to incorporate activity information from other agencies, when appropriate, in their overall assessments.

A second, related issue stems from the different reporting procedures that organizations use to collect and evaluate AIA data. How information is documented at different levels and offices across an organization affects the composition of its overall AIA assessment. In some commands, an appropriate and well-developed AIA reporting chain appears to exist, whereas reporting relationships in other organizations are less logical, clear, or consistent. In the process of implementing AIAKSS, HQDA should work with its MACOMs to ensure that AIA officials at the command level have identified a clear method for reporting field-level data to a higher level where it can be properly aggregated and evaluated.

Third, there is the macro-micro problem: How should the commands aggregate data on individual Army International Activities, such as multinational exercises, in support of larger Army goals, such as operational access?[3] Certain commands have AIA program managers who understand the rationale of the activities they support and are cognizant of their results. They would seem to be the natural candidates for converting AIA outputs into outcome evaluations. However, not all international activities are linked to a program manager within an Army command. Furthermore, not all managers are equally knowledgeable regarding the outcomes of their programs and the challenges they face in the field. One solution to this dilemma is for each command, in cooperation with HQDA, to assign one or more experienced and able officials the job of aggregating the assessment results of every AIA program for which that command is in some way responsible. Another alternative is for each command to compile information on programs for which it is primarily responsible and can verify the results.

Finally, there is the issue of program versus country evaluations. The regional Combatant Commands are building assessment tools designed to provide information on the contribution that security cooperation activities are making toward establishing or maintaining new relationships, capabilities, and access in DoD priority countries. AIAKSS, on the other hand, is primarily focused on demonstrating the performance of international programs in meeting U.S. Army and national security objectives. Both approaches are equally valid and important, but some mechanism must be created for reconciling country and program approaches to assessment. We suggest that HQDA work with OSD, the Joint Staff, the regional Combatant Commands and other agencies inside and outside DoD to create a U.S. government security cooperation assessment framework that integrates intermediate assessment mechanisms through a common set of ends and ways as well as a capability for sharing and using information on international activities regardless of where they occur and to whom they belong.

---

[3] This phenomenon also afflicts social science research. How individual acts produce particular group behavior has long been a puzzle among social scientists. For more on this subject, see Schelling (1978).

Appendix: AIA Performance Indicators

Table A.1
AIA Performance Indicators

	Output Indicators	Outcome Indicators
	AIA Way: Education and Training	
1. Ensure access	1. Number of courses that addressed issues synonymous with or supportive of access	1. Education and training contributed directly to expand access—in type—to foreign assets
	2. % change over previous year in number of courses that addressed issues synonymous with or supportive of access	2. Education and training contributed directly to expand access—in intensity—to foreign assets
	3. Number of activities that addressed issues synonymous with or supportive of access	3. U.S. "graduates" were placed in positions to support access
	4. % change over previous year in number of activities that addressed issues synonymous with or supportive of access	4. Foreign "graduates" were placed in positions to support access
	5. Number of U.S. personnel involved in these courses and activities	5. Formal and informal networks to sustain contact between U.S. and foreign "graduates" were leveraged to expand access
	6. % change over previous year in number of U.S. personnel involved in these courses and activities	
	7. Number of foreign personnel involved in these courses and activities	
	8. % change over previous year in number of foreign personnel involved in these courses and activities	
2. Promote transformation	1. Number of courses that addressed issues synonymous with or supportive of transformation	1. Formal and informal networks that sustain contact between U.S. and foreign "graduates" were leveraged to support transformation
	2. % change over previous year in number of courses that addressed issues synonymous with or supportive of transformation	2. New formal and informal networks were formed to link U.S. and foreign "graduates" to support transformation

Table A.1 (continued)

	Output Indicators	Outcome Indicators
	3. Number of activities that addressed issues synonymous with or supportive of transformation 4. % change over previous year in number of activities that addressed issues synonymous with or supportive of transformation	3. Education and training expanded U.S. capacity to support transformation 4. Education and training expanded foreign capacity to work with transformation
3. Improve interoperability	1. Number of courses supporting interoperability 2. % change over previous year in number of courses supporting interoperability 3. Number of billets offered to foreign personnel for courses supporting interoperability 4. % change over previous year in number of billets offered to foreign personnel for courses supporting interoperability 5. Number of foreign countries sending personnel to courses supporting interoperability 6. % change over previous year in number of foreign countries sending personnel to courses supporting interoperability 7. Number of long-term agreements/exchanges for education and training in interoperability 8. % change over previous year in number of long-term agreements/exchanges for education and training in interoperability 9. Number of certification agreements with foreign army education and training institutions 10. % change in number of certification agreements with foreign army education and training institutions	1. Education and training contributed directly to expand U.S. capacity to close or reduce critical gaps in interoperability 2. Education and training contributed directly to expand foreign capacity to close or reduce critical gaps in interoperability 3. Formal and informal networks that sustain contact between U.S. and foreign "graduates" were leveraged to promote interoperability 4. New formal or informal networks were formed to sustain ties between U.S. and foreign "graduates" to promote interoperability

Table A.1 (continued)

	Output Indicators	Outcome Indicators
4. Improve defense capabilities	1. Number of billets assigned to non-traditional allies 2. % change over previous year in number of billets assigned to non-traditional allies	1. Expanded foreign defense capabilities to deal with domestic threats and humanitarian relief 2. Expanded foreign defense capabilities to deal with external threats 3. Expanded foreign defense capabilities to contribute to regional security 4. Expanded foreign defense capabilities in bilateral defense relations with the United States
5. Promote stability and democracy	1. Number of billets assigned to non-traditional NATO allies 2. % change over previous year in number of billets assigned to non-traditional NATO allies 3. Number of courses that addressed issues synonymous with or supportive of democracy and stability, including good governance, accountability, institution building, etc. 4. % change over previous year in number of courses that addressed issues synonymous with or supportive of democracy and stability 5. Number of activities that addressed issues synonymous with or supportive of democracy and stability 6. % in number of activities synonymous with or supportive of democracy and stability	1. Formal and informal networks that sustain contact between U.S. and foreign "graduates" were leveraged to promote democracy and stability 2. New formal or informal networks were formed to sustain ties between U.S. and foreign "graduates" to promote democracy and stability

Table A.1 (continued)

	Output Indicators	Outcome Indicators
6. Assure allies	1. Total number of billets assigned to U.S. personnel 2. % change over previous year in total number of billets assigned to U.S. personnel 3. Total number of billets assigned to traditional allies 4. % change over previous year in total number of billets assigned to traditional allies 5. Total number of billets assigned to non-traditional allies 6. % change over previous year in total number of billets assigned to non-traditional allies 7. Number of courses or activities that addressed issues of critical concern to participating foreign countries 8. % change in number of courses or activities that addressed issues of critical concern to participating foreign countries 9. Average time it took to process requests for education and training from foreign countries 10. % change over previous year in average time it took to process requests for education and training from foreign countries 11. Amount of grants allocated to foreign countries participating in education and training 12. % change over previous year in amount of grants allocated to foreign countries participating in education and training	1. Formal and informal networks to sustain contact between U.S. and foreign "graduates" were leveraged to support bilateral defense relations 2. New formal or informal networks were formed to sustain ties between U.S. and foreign "graduates" to support bilateral defense relations

Table A.1 (continued)

	Output Indicators	Outcome Indicators
7. Improve non-military cooperation	1. Number of courses that addressed issues synonymous with or supportive of cooperation with the United States 2. % change over previous year in number of courses that addressed issues synonymous with or supportive of cooperation with the United States 3. Number of activities that addressed issues synonymous with or supportive of cooperation with the United States 4. % change over previous year in number of activities that addressed issues synonymous with or supportive of cooperation with the United States 5. Number of U.S. personnel who took part in education and training courses and activities overseas 6. % change over previous year in number of U.S. personnel who took part in education and training courses and activities overseas 7. Number of foreign personnel from traditional allies received in the United States for Army education and training 8. % change over previous year in number of foreign personnel from traditional allies received in the United States for Army education and training 9. Number of foreign personnel from non-traditional allies received in the United States for Army education and training 10. % change over previous year in number of foreign personnel from non-traditional allies received in the United States for Army education and training	1. Formal and informal networks to sustain contact between U.S. and foreign "graduates" were leveraged to support bilateral defense cooperation 2. New formal and informal networks were formed to sustain ties between U.S. and foreign "graduates" to support bilateral defense cooperation

Table A.1 (continued)

Output Indicators	Outcome Indicators
8. Establish relations	
1. Number of foreign countries that were first-time participants in Army education and training courses and activities	1. Opened new foreign institutional contact for Army education and training
2. % change over previous year in number of first-time foreign countries in Army education and training courses and activities	2. Links to new formal and information networks for U.S. and foreign "graduates" were created
3. Number of foreign country personnel who were first-time participants in Army education and training courses and activities	
4. % change over previous year in number of first-time foreign personnel in Army education and training courses and activities	

AIA Way: International Support Arrangement (ISA) and Treaty Compliance (TC)

Output Indicators	Outcome Indicators
1. Ensure access	
1. Concluded new access agreements or arrangements	1. U.S. gained expanded access—in type—to foreign assets
2. Total number of ISA/TC events and activities that explicitly emphasized access for the United States	2. U.S. gained expanded access—in intensity—to foreign assets
3. % change over previous year in the number of ISA/TC events and activities that explicitly emphasized access for the United States	3. Formal and informal networks associated with ISA and TC were leveraged to support access
2. Promote transformation	
1. Number of ISA/TC events or activities explicitly supportive of transformation	1. Events or activities enabled the United States to acquire assets or expand capacity for transformation
2. % change over previous year in the number of ISA/TC events or activities explicitly supportive of transformation	2. Formal and informal networks associated with ISA and TC were leveraged to support transformation

Table A.1 (continued)

	Output Indicators	Outcome Indicators
3. Improve interoperability	1. Number of ISA/TC events or activities explicitly supportive of interoperability 2. % change over previous year in the number of ISA/TC events or activities explicitly supportive of interoperability	1. Events or activities enabled foreign countries to close or reduce critical gaps in interoperability 2. Events or activities enabled the United States to close or reduce critical gaps in interoperability 3. Formal and informal networks associated with ISA and TC were leveraged to support interoperability
4. Improve defense capabilities	1. Number of ISA/TC events or activities that explicitly aimed to improve foreign defense capabilities 2. % change over previous years in number of ISA/TC events or activities that explicitly aimed to improve foreign defense capabilities	1. Expanded foreign defense capabilities to deal with domestic threats and humanitarian relief 2. Expanded foreign defense capabilities to deal with external threats 3. Expanded foreign defense capabilities to contribute to regional security 4. Expanded foreign defense capabilities in bilateral defense relations with the United States
5. Promote stability and democracy	1. Number of ISA/TC events or activities that explicitly addressed democracy and stability, including good governance, accountability, institution building, etc. 2. % change over the previous year in the number of ISA/TC events or activities that explicitly addressed democracy and stability 3. Number of countries that participated in ISA/TC events or activities that explicitly addressed democracy and stability 4. % change over the previous year in the number of ISA/TC events or activities that explicitly addressed democracy and stability	1. Formal and informal networks associated with ISA and TC were leveraged to support democracy and stability

Table A.1 (continued)

	Output Indicators	Outcome Indicators
6. Assure allies	1. Current level of U.S. staffing to multilateral enterprise 2. % change over previous year in U.S. staffing level to multilateral enterprise 3. Number of ISA/TC events or activities the United States took part in 4. % change over previous year in the number of ISA/TC events or activities the United States took part in 5. Number of ISA/TC events or activities led or organized by the United States 6. % change over previous year in the number of ISA/TC events or activities led or organized by the United States	1. Number of countries that took part in ISA/TC events or activities led or organized by the United States 2. % change over previous year the number of countries that took part in ISA/TC events or activities led or organized by the United States 3. Formal and informal networks associated with ISA and TC were leveraged to support bilateral defense relations
7. Improve non-military cooperation	1. Current level of U.S. staffing to multilateral enterprise 2. % change over previous year in U.S. staffing level to multilateral enterprise 3. Number of ISA/TC events or activities the United States took part in 4. % change over previous year in the number of ISA/TC events or activities the United States took part in 5. Number of events or activities led or organized by the United States 6. % change over previous year in the number of events or activities led or organized by the United States	1. Number of countries that took part in ISA/TC events or activities led or organized by the United States 2. % change over previous year the number of countries that took part in ISA/TC events or activities led or organized by the United States 3. Formal and informal networks associated with ISA and TC were leveraged to support bilateral defense relations
8. Establish relations	1. Number of countries involved in ISA and TC for the first time 2. % change in number of countries involved in ISA and TC for the first time	1. Opened new foreign institutional contact for Army ISA and TC

Table A.1 (continued)

	Output Indicators	Outcome Indicators
	AIA Way: Materiel Transfer (MT) and Technical Training (TT)	
1. Ensure access	1. Number of foreign countries that received MT and TT from the United States 2. % change over previous year in number of foreign countries that received MT and TT from the United States 3. Number of high-technology niche equipment or technology transferred to the United States 4. % change over previous year in number of high-technology niche equipment or technology transferred to the United States	1. MT and TT helped the United States to gain expanded access—in type—to foreign assets 2. MT and TT helped the United States to gain expanded access—in intensity—to foreign assets 3. Formal and informal networks associated with MT and TT were leveraged to expand access
2. Promote transformation	1. Number of high-technology niche equipment or technology transfer to the United States 2. % change over previous year in number of high-technology niche equipment or technology transfer to the United States 3. Number of MT and TT events approved for foreign countries 4. % change over previous year in number of MT and TT events approved for foreign countries 5. Number of MT and TT events approved for traditional allies 6. % change over previous year in number of MT and TT events approved for traditional allies 7. Number of MT and TT events approved for non-traditional allies 8. % change over previous year in number of MT and TT events approved for non-traditional allies 9. Number of foreign countries that received MT and TT from the United States 10. % change over previous year in number of foreign countries that received MT and TT from the United States	1. MT and TT expanded U.S. capacity to support transformation 2. MT and TT expanded capacity for foreign countries to work with transformation 3. Formal and informal networks associated with MT and TT were leveraged to support transformation

Table A.1 (continued)

	Output Indicators	Outcome Indicators
3. Improve interoperability	1. Number of foreign countries that received MT and TT 2. % change over previous year in number of foreign countries that received MT and TT 3. Number of traditional allies that received MT and TT 4. % change over previous year in number of traditional allies that received MT and TT 5. Number of non-traditional allies that received MT and TT 6. % change over previous year in number of non-traditional allies that received MT and TT 7. Number of MT and TT with traditional allies with stated emphasis to boost interoperability 8. % change over previous year in number of MT and TT with traditional allies with stated emphasis to boost interoperability 9. Number of MT and TT with non-traditional allies with stated emphasis on boosting interoperability 10. % change over previous year in number of MT and TT with non-traditional allies with stated emphasis on boosting interoperability	1. MT and TT directly helped to expand U.S. capacity to close or reduce critical gaps in interoperability 2. MT and TT directly helped to expand foreign capacity to close or reduce critical gaps in interoperability 3. Number of countries using U.S. defense equipment as platform for modernization 4. % change over previous year in the number of countries using U.S. defense equipment as platform for modernization 5. Formal and informal networks associated with MT and TT were leveraged to support interoperability
4. Improve defense capabilities	1. Number of MT and TT events approved for foreign countries 2. % change over previous year in number of MT and TT events approved for foreign countries 3. Number of foreign countries that received MT and TT from the United States 4. % change over previous year in number of countries that received MT and TT from the United States	1. Expanded foreign defense capabilities to deal with domestic threats and humanitarian relief 2. Expanded foreign defense capabilities to deal with external threats 3. Expanded foreign defense capabilities to contribute to regional security 4. Expanded foreign defense capabilities in bilateral defense relations with the United States

Table A.1 (continued)

	Output Indicators	Outcome Indicators
5. Promote stability and democracy	1. Number of non-traditional NATO countries that received MT and TT from the United States 2. % change over previous year in number of non-traditional NATO countries that received MT and TT from the United States	1. MT and TT recipient countries complied with U.S. anti-bribery laws 2. Formal and informal networks associated with MT and TT were leveraged to support transformation
6. Assure allies	1. Value of MT and TT transferred from the United States 2. % change over previous year in value of MT and TT transferred from the United States 3. Number of MT and TT events approved for foreign countries 4. % change over previous year in number of MT and TT events approved for foreign countries 5. Number of foreign countries that received MT and TT from the United States 6. % change over previous year in number of countries that received MT and TT from the United States 7. Number of sales contracts 8. % change over previous year in number of sales contracts 9. Number of leases 10. % change over previous year in number of leases 11. Number of requests for MT and TT received 12. % change over previous year in number of requests for MT and TT received 13. Number of requests for MT and TT received from traditional allies 14. % change over previous year in number of requests for MT and TT received from traditional allies	1. Positive feedback received from customer satisfaction surveys for U.S. Army contracts support 2. Number of foreign countries using U.S. defense equipment as platform for modernization 3. % change over previous year in the number of foreign countries using U.S. defense equipment as platform for modernization 4. Number of traditional allies using U.S. defense equipment as platform for modernization 5. % change over previous year in the number of traditional allies using U.S. defense equipment as platform for modernization 6. Number of non-traditional allies using U.S. defense equipment as platform for modernization 7. % change over previous year in the number of non-traditional allies using U.S. defense equipment as platform for modernization 8. Formal and informal networks associated with MT and TT were leveraged to support bilateral defense relations

Table A.1 (continued)

	Output Indicators	Outcome Indicators
	15. Number of requests for MT and TT received from non-traditional allies	
	16. % change over previous year in number of requests for MT and TT received from non-traditional allies	
	17. Total number of requests for MT and TT granted	
	18. % change over previous year in total number of requests for MT and TT granted	
	19. Number of requests for MT and TT granted to traditional allies	
	20. % change over previous year in number of requests for MT and TT granted to traditional allies	
	21. Number of requests for MT and TT granted to non-traditional allies	
	22. % change over previous year in number of requests for MT and TT granted to non-traditional allies	
7. Improve non-military cooperation	1. Value of MT and TT transferred from the United States	1. Number of foreign countries using U.S. defense equipment as platform for modernization
	2. % change over previous year in value of MT and TT transferred from the United States	2. % change over previous year in the number of foreign countries using U.S. defense equipment as platform for modernization
	3. Number of MT and TT events approved for foreign countries	3. Number of traditional allies using U.S. defense equipment as platform for modernization
	4. % change over previous year in number of MT and TT events approved for foreign countries	4. % change over previous year in the number of traditional allies using U.S. defense equipment as platform for modernization
	5. Number of MT and TT events approved for traditional allies	5. Number of non-traditional allies using U.S. defense equipment as platform for modernization
	6. % change over previous year in number of MT and TT events approved for traditional allies	6. % change over previous year in the number of non-traditional allies using U.S. defense equipment as platform for modernization
	7. Number of MT and TT events approved for non-traditional allies	

Table A.1 (continued)

	Output Indicators	Outcome Indicators
	8. % change over previous year in number of MT and TT events approved for non-traditional allies	7. Formal and informal networks associated with MT and TT were leveraged to support bilateral defense relations
	9. Number of sales contracts and leases	
	10. % change over previous year in number of sales contracts and leases	
	11. Number of sales contracts and leases with traditional allies	
	12. % change over previous year in number of sales contracts and leases with traditional allies	
	13. Number of sales contracts and leases with non-traditional allies	
	14. % change over previous year in number of sales contracts and leases with non-traditional allies	
	15. Number of requests for MT and TT received	
	16. % change over previous year in number of requests for MT and TT received	
	17. Number of requests for MT and TT received from traditional allies	
	18. % change over previous year in number of requests for MT and TT received from traditional allies	
	19. Number of requests for MT and TT received from non-traditional allies	
	20. % change over previous year in number of requests for MT and TT received from non-traditional allies	
8. Establish relations	1. Number of first time MT and TT requests received	1. Opened new foreign institutional contacts for Army MT and TT
	2. % change over previous year in number of first time MT and TT requests received	

Table A.1 (continued)

	Output Indicators	Outcome Indicators
	AIA Way: Military-to-Military Contacts (MMC)	
1. Ensure access	1. Number of MMC 2. % change over previous year in number of MMC 3. Number of MMC that made access a stated objective 4. % change over previous year in number of MMC that made access a stated objective	1. MMC helped the United States to gain expanded access—in type—to foreign assets 2. MMC helped the United States to gain expanded access—in intensity—to foreign assets 3. Formal and informal networks associated with MMC were leveraged to expand access
2. Promote transformation	1. Number of MMC that made transformation a stated objective 2. % change over previous year in number of MMC that made transformation a stated objective	1. MMC contributed to high-level agreement to support transformation 2. MMC contributed to increase foreign capacity to work with transformation 3. Formal and informal networks associated with MMC were leveraged to support transformation
3. Improve interoperability	1. Number of MMC that made interoperability a stated objective 2. % change over previous year in number of MMC that made interoperability a stated objective 3. Number of MMC with traditional allies 4. % change over previous year in number of MMC with traditional allies 5. Number of MMC with non-traditional allies 6. % change over previous year in number of MCC with non-traditional allies	1. MMC contributed to high-level agreement to increase interoperability 2. MMC contributed to high-level agreement to expand capacity of foreign country to increase interoperability 3. Formal and informal networks associated with MMC were leveraged to support interoperability

Table A.1 (continued)

	Output Indicators	Outcome Indicators
4. Improve defense capabilities	1. Number of MMC that made improving defense capability a stated goal of interaction with foreign personnel 2. % change over previous year in the number of MMC that made improving defense capability a stated goal of interaction with foreign personnel 3. Number of MMC with traditional allies 4. % change over previous year in number of MMC with traditional allies 5. Number of MMC with non-traditional allies 6. % change over previous year in number of MMC with non-traditional allies	1. MMC directly responded to foreign request for assistance to expand defense capability 2. MMC directly paved way for foreign country to acquire necessary U.S. assistance to improve its defense capabilities 3. Formal and informal networks associated with MMC were leveraged to promote foreign defense capabilities
5. Promote stability and democracy	1. Number of MMC that explicitly addressed topics synonymous with or supportive of democracy and stability, including good governance, accountability, institution building, etc. 2. % change over previous year in number of MMC that explicitly addressed topics synonymous with or supportive of democracy and stability	1. Formal and informal networks associated with MMC were leveraged to support interoperability
6. Assure allies	1. Total number of MMC 2. % change over previous year in total number of MMC 3. Number of MMC with traditional allies 4. % change over previous year in number of MMC with traditional allies 5. Number of MMC with non-traditional allies 6. % change over previous year in number of MMC with non-traditional allies	1. Formal and informal networks associated with MMC were leveraged to support bilateral defense relations with traditional allies 2. Formal and informal networks associated with MMC were leveraged to support bilateral defense relations with traditional allies 3. Formal and informal networks associated with MMC were leveraged to support bilateral defense relations with non-traditional allies

Table A.1 (continued)

	Output Indicators	Outcome Indicators
	7. Number of U.S. attachés and Foreign Liaison Officers (FLOs) assigned to foreign countries 8. % change over previous year in number of U.S. attachés and FLOs assigned to foreign countries 9. Number of foreign attachés and FLOs assigned to the United States 10. % change over previous year in number of foreign attachés and FLOs assigned to the United States	4. Expanded Army contact with senior officials of traditional allies 5. Expanded Army contact with senior officials of non-traditional allies
7. Improve non-military cooperation	1. Number of U.S. attachés and FLOs assigned to foreign countries 2. % change over previous year in number of U.S. attachés and FLOs assigned to foreign countries 3. Number of foreign attachés and FLOs assigned to the United States 4. % change over previous year in number of foreign attachés and FLOs assigned to the United States 5. Total number of MMC 6. % change over previous year in total number of MMC 7. Number of MMC with traditional allies 8. % change over previous year in number of MMC with traditional allies 9. Number of MMC with non-traditional allies 10. % change over previous year in number of MMC with non-traditional allies	1. Formal and informal networks associated with MMC were leveraged to support bilateral defense relations 2. Formal and informal networks associated with MMC were leveraged to support bilateral defense relations with traditional allies 3. Formal and informal networks associated with MMC were leveraged to support bilateral defense relations with non-traditional allies 4. Expanded Army contact with senior officials of traditional allies 5. Expanded Army contact with senior officials of non-traditional allies
8. Establish relations	1. Formal agreement signed to conduct MMC	1. Opened new foreign institutional contacts for Army senior officials

Table A.1 (continued)

	Output Indicators	Outcome Indicators
	AIA Way: Military-to-Military Exchanges (MMEx)	
1. Ensure access	1. Number of MMEx that involved U.S. personnel primarily active in access-related work 2. % change over previous year in number of MMEx that involved U.S. personnel primarily active in access-related work 3. Number of MMEx that supported access for the United States 4. % change over previous year in number of MMEx that supported access for the United States	1. MMEx helped the United States to gain expanded access—in type—to foreign assets 2. MMEx helped the United States to gain expanded access—in intensity—to foreign assets 3. Formal and informal networks associated with MMEx were leveraged to expand access 4. U.S. personnel were assigned to support access work following MMEx 5. Foreign personnel were assigned to support access work following MMEx
2. Promote transformation	1. MMEx facilitated U.S. access to technology and other assets 2. Number of MMEx with stated goal of supporting transformation 3. % change in number of MMEx with stated goal of supporting transformation 4. Number of MMEx with traditional allies 5. % change over previous year in number of MMEx with traditional allies 6. Number of MMEx with non-traditional allies 7. % change over previous year in number of MMEx with non-traditional allies 8. Number of MMEx that explicitly targeted acquisition of technology and other assets to support transformation 9. % change over previous year in number of MMEx that explicitly targeted acquisition of technology and other assets to support transformation	1. Formal and informal networks associated with MMEx were leveraged to support transformation 2. MMEx expanded U.S. capacity to support transformation 3. MMEx expanded foreign capacity to work with transformation

Table A.1 (continued)

	Output Indicators	Outcome Indicators
3. Improve interoperability	1. Number of MMEx with stated goal of improving interoperability 2. % change over previous year in number of MMEx with stated goal of improving interoperability 3. Number of personnel in MMEx who support interoperability 4. % change over previous year in number of personnel in MMEx who support interoperability	1. MMEx directly enabled the United States to close or reduce critical gaps in interoperability 2. MMEx directly enabled allies to close or reduce critical gaps in interoperability with the United States 3. Formal and informal networks associated with MMEx were leveraged to support interoperability
4. Improve defense capabilities	1. MMEx emphasized or had stated goal of improving defense capabilities of foreign countries 2. Number of MMC with non-traditional allies 3. % change over previous year in number of MMC with non-traditional allies	1. Expanded foreign defense capabilities to deal with domestic threats and humanitarian relief 2. Expanded foreign defense capabilities to deal with external threats 3. Expanded foreign defense capabilities to contribute to regional security 4. Expanded foreign defense capabilities in bilateral defense relations with the United States
5. Promote stability and democracy	1. Number of MMEx that emphasized civil-military relations 2. % change over previous year in number of MMEx that emphasized civil-military relations 3. Number of foreign MMEx personnel with responsibilities in civil-military relations 4. % change over previous year in number of foreign MMEx personnel with responsibilities in civil-military relations 5. Number of MMEx with non-traditional allies 6. % change over previous year in number of MMEx with non-traditional allies	1. Formal and informal networks associated with MMEx were leveraged to support democracy and stability, including good governance, accountability, institution building, etc. 2. MMEx contributed to expand democracy and stability in foreign country

Table A.1 (continued)

	Output Indicators	Outcome Indicators
6. Assure allies	1. Number of MMEx with all countries 2. % change over previous year in total number of MMEx 3. Number of MMEx with traditional allies 4. % change over previous year in number of MMEx with traditional allies 5. Number of MMEx with non-traditional allies 6. % change over previous year in number of MMEx with non-traditional allies 7. Number of existing long-term exchange agreements 8. % change over previous year in number of long-term exchange agreements 9. Number of new long-term exchange agreements concluded 10. % change over previous year in number of new long-term exchange agreements concluded	1. Formal and informal networks associated with MMEx were leveraged to support bilateral defense relations 2. MMEx contributed to deepening mutual assurance
7. Improve non-military cooperation	1. Number of MMEx with all countries 2. % change over previous year in number of MMEx with all countries 3. Number of MMEx with traditional allies 4. % change over previous year in number of MMEx with traditional allies 5. Number of MMEx with non-traditional allies 6. % change over previous year in number of MMEx with non-traditional allies 7. Number of existing long-term exchange agreements 8. % change over previous year umber of long-term exchange agreements	1. Formal and informal networks associated with MMEx were leveraged to support bilateral defense relations 2. MMEx contributed to deepening bilateral defense cooperation
8. Establish relations	1. Number of new long-term exchange agreements signed 2. % change over previous year in number of long-term exchange agreements signed	1. MMEx opened new institutional point of contact for the Army

Table A.1 (continued)

	Output Indicators	Outcome Indicators
	AIA Way: Military Exercises (ME)	
1. Ensure access	1. Number of existing long-term access agreements 2. % change over previous year in number of existing long-term access agreements 3. Number of ME that had access to physical facilities overseas 4. % change over previous year in number of ME that had access to physical facilities overseas	1. ME helped the United States to expand access—in type—to foreign assets 2. ME helped the United States to expand access—in intensity—to foreign assets 3. Formal and informal networks associated with ME were leveraged to support access
2. Promote transformation	1. Total number of ME with explicit emphasis to support transformation 2. % change over previous year in number of ME with explicit emphasis to support transformation	1. ME enabled the United States to acquire foreign technologies and other assets supportive of transformation 2. ME enabled validation of the applicability and effectiveness of new technologies and approaches to warfighting under transformation 3. Formal and informal networks associated with ME were leveraged to support transformation 4. Expanded U.S. capacity to support interoperability 5. Expanded foreign capacity to interoperate with the United States
3. Improve interoperability	1. Number of ME conducted 2. % change in number of ME conducted 3. Number of multilateral agreements that set common standards and concepts of operation	1. ME directly expanded U.S. capacity to close or reduce critical gaps in interoperability 2. ME directly expanded foreign capacity to close or reduce gaps in interoperability with the United States 3. ME tested applicability and effectiveness of new technology and approaches to warfighting

Table A.1 (continued)

	Output Indicators	Outcome Indicators
	4. % change over previous year in number of multilateral agreements that set common standards and concepts of operation 5. Number of foreign countries adopting common standards and concepts of operation 6. % change over previous year in number of foreign countries adopting common standards and concepts of operation	4. Formal and informal networks associated with ME were leveraged to support interoperability
4. Improve defense capabilities	1. Number of small unit exercises that introduced technology and equipment to foreign countries to build capacity to participate in ME 2. % change over previous year in number of small unit exercises that introduced technology and equipment to foreign countries to build capacity to participate in ME 3. Number of activities that trained foreign countries in multilateral ME planning and execution 4. % change over previous year in number of activities that trained foreign countries in multilateral ME planning and execution 5. Number of ME with non-traditional allies 6. % change over previous year in number of ME with non-traditional allies	1. Foreign country gained capability to work in multilateral ME 2. Expanded foreign defense capabilities to deal with domestic threats and humanitarian relief 3. Expanded foreign defense capabilities to deal with external threats 4. Expanded foreign defense capabilities to contribute to regional security 5. Expanded foreign defense capabilities in bilateral defense relations with the United States

Table A.1 (continued)

	Output Indicators	Outcome Indicators
5. Promote stability and democracy	1. Number of foreign civilian observers at ME 2. % change over previous year in number of foreign civilian observers at ME 3. Number of ME with explicit emphasis on fostering civil-military relations 4. % change over previous year in number of ME with explicit emphasis on fostering civil-military relations 5. Number of ME with non-traditional allies 6. % change over previous year in number of ME with non-traditional allies	1. Formal and informal networks associated with ME were leveraged to support democracy and stability
6. Assure allies	1. Number of ME activities with all countries 2. % change over previous year in number of ME activities with all countries 3. Number of foreign countries participating in ME activities 4. % change in number of foreign countries participating in ME activities 5. Number of ME with traditional allies 6. % change over previous year in number of ME with traditional allies 7. Number of ME with non-traditional allies 8. % change over previous year in number of ME with non-traditional allies 9. Number of ME activities conducted in foreign countries 10. % change in number of ME activities conducted in foreign countries	1. Formal and informal networks associated with ME were leveraged to support bilateral defense relations

Table A.1 (continued)

	Output Indicators	Outcome Indicators
7. Improve non-military cooperation	1. Number of ME activities with all countries 2. % change over previous year in number of ME activities with all countries 3. Number of foreign countries participating in ME activities 4. % change in number of foreign countries participating in ME activities 5. Number of ME activities conducted in foreign countries 6. % change in number of ME activities conducted in foreign countries	1. Formal and informal networks associated with ME were leveraged to support bilateral defense relations
8. Establish relations	1. The United States used ME activities to expand contacts with foreign countries	1. Opened new foreign institutional contact for Army
AIA Way: RDT&E		
1. Ensure access	1. Number of countries participating in RDT&E activities with the United States 2. % change over previous year in number of countries participating in RDT&E activities with the United States 3. Number of RDT&E activities with traditional allies 4. % change over previous year in number of RDT&E activities with traditional allies 5. Number of RDT&E activities with non-traditional allies 6. % change over previous year in number of RDT&E activities with non-traditional allies	1. RDT&E activities enabled the United States to access new technologies, research data and facilities, etc.

Table A.1 (continued)

	Output Indicators	Outcome Indicators
2. Promote transformation	1. Number of countries participating in RDT&E activities with the United States 2. % change over previous year in number of countries participating in RDT&E activities with the United States 3. Number of RDT&E activities that emphasized support for transformation 4. % change in number of RDT&E activities that emphasized support for transformation 5. Number of RDT&E activities with traditional allies 6. % change over previous year in number of RDT&E activities with traditional allies 7. Number of RDT&E activities with non-traditional allies 8. % change over previous year in number of RDT&E activities with non-traditional allies	1. RDT&E activities enabled the United States to access new technologies, research data, facilities, etc., supportive of transformation 2. Expanded U.S. capacity to support transformation 3. Expanded foreign capacity to work with transformation
3. Improve interoperability	1. Number of countries participating in RDT&E activities with the United States 2. % change over previous year in number of countries participating in RDT&E activities with the United States 3. Number of RDT&E activities that explicitly emphasized improving interoperability 4. % change in number of RDT&E activities that explicitly emphasized improving interoperability	1. RDT&E activities expanded U.S. capacity to close or reduce critical gaps in interoperability 2. RDT&E activities expanded foreign capacity to close or reduce critical gaps in interoperability with the United States 3. Formal and informal networks associated with RDT&E were leveraged to support interoperability

Table A.1 (continued)

	Output Indicators	Outcome Indicators
4. Improve defense capabilities	1. New RDT&E activities conducted with foreign countries seeking assistance to bolster basic research capability 2. Number of RDT&E activities with non-traditional allies 3. % change over previous year in number of RDT&E activities with non-traditional allies	1. RDT&E activities expanded basic research capability in foreign country
5. Promote stability and democracy	None articulated	None articulated
6. Assure allies	1. Number of RDT&E activities with all countries 2. % change over previous year in number of RDT&E activities with all countries 3. Number of U.S. personnel involved in RDT&E activities 4. % change over previous year in number of U.S. personnel involved in RDT&E activities 5. Number of U.S. personnel involved in RDT&E activities with traditional allies 6. % change over previous year in number of U.S. personnel involved in RDT&E activities with traditional allies 7. Number of U.S. personnel involved in RDT&E activities with non-traditional allies 8. % change over previous year in number of U.S. personnel involved in RDT&E activities with non-traditional allies 9. Number of RDT&E activities hosted in the United States	1. RDT&E activities expanded formal and informal networks that provide mechanisms and channels to address issues in bilateral defense relations 2. Formal and informal networks associated with RDT&E were leveraged to support bilateral defense relations

Table A.1 (continued)

	Output Indicators	Outcome Indicators
	10. % change over previous year in number of RDT&E activities hosted in the United States 11. Number of RDT&E activities hosted in foreign countries 12. % change over previous year in number of RDT&E activities hosted in foreign countries	
7. Improve non-military cooperation	1. Number of agreements to exchange sensitive data, share RDT&E facilities, conduct personnel exchange, etc. 2. % change over previous year in number of agreements to exchange sensitive data, share RDT&E facilities, conduct personnel exchange, etc. 3. Number of RDT&E activities with foreign countries 4. % change over previous year in number of RDT&E activities with foreign countries 5. Number of U.S. personnel involved in RDT&E activities 6. % change over previous year in number of U.S. personnel involved in RDT&E activities 7. Number of foreign personnel involved in RDT&E activities with the United States 8. % change over previous year in number of foreign personnel involved in RDT&E activities with the United States 9. Number of foreign countries involved in RDT&E activities with the United States 10. % change in number of foreign countries involved in RDT&E activities with the United States	1. Formal and informal networks associated with ME were leveraged to support bilateral defense relations
8. Establish relations	1. New agreements signed to exchange sensitive data, share RDT&E facilities, conduct personnel exchange, etc.	1. Opened new foreign institutional contact for Army RDT&E

Table A.1 (continued)

Output Indicators	Outcome Indicators
AIA Way: Standing Forums	
1. Ensure access	1. Number of meetings with explicit objective of addressing access issues 2. % change over previous year in number of meetings with explicit objective of addressing access issues	1. Forum meetings and activities expanded formal and informal networks that provide mechanisms and channels to address issues in access 2. Formal and informal networks associated with ME were leveraged to support bilateral defense relations 3. Forum meetings and activities expanded access—in type—to foreign assets 4. Forum meetings and activities expanded access—in intensity—to foreign assets
2. Promote transformation	1. Number of meetings with explicit emphasis on transformation issues 2. % change over previous year in number of meetings with explicit emphasis on transformation issues 3. Number of permanent or quasi-permanent working groups to address transformation-related issues 4. % change over previous year in number of permanent or quasi-permanent working groups to address transformation-related issues 5. Number of new agreements concluded to promote transformation-related objectives 6. % change over previous year in number of new agreements concluded to promote transformation-related objectives	1. Formal and informal networks associated with standing forums were leveraged to support transformation 2. Forum meetings and activities expanded formal and informal networks that provide mechanisms and channels to address issues in transformation

Table A.1 (continued)

	Output Indicators	Outcome Indicators
3. Improve interoperability	1. Number of forum participants involved in interoperability-related discussions 2. % change over previous year in number of forum participants involved in interoperability-related discussions 3. Number of forum meetings addressing interoperability 4. % change over previous year in number of forum meetings addressing interoperability 5. Number of forum meeting with explicit emphasis on improving interoperability 6. % change over previous year in number of forum meeting with explicit emphasis on improving interoperability	1. Formal and informal networks associated with standing forums were leveraged to support interoperability 2. Forum meetings and activities expanded formal and informal networks that provide mechanisms and channels to address issues in interoperability
4. Improve defense capabilities	1. Number of forum meetings and activities that explicitly emphasized assistance to improve defense capabilities of forum members 2. % change in number of forum meetings and activities that explicitly emphasized assistance to improve defense capabilities of forum members 3. Number of forum meetings and activities that explicitly emphasized assistance to improve defense capabilities of non-forum members 4. % change over previous year in number of forum meetings and activities that explicitly emphasized assistance to improve defense capabilities of non-forum members	1. Forum meetings and activities directly contributed to bolster defense capabilities of forum members 2. Forum meetings and activities directly contributed to bolster defense capabilities of non-forum members 3. Expanded foreign defense capabilities to deal with domestic threats and humanitarian relief 4. Expanded foreign defense capabilities to deal with external threats 5. Expanded foreign defense capabilities to contribute to regional security 6. Expanded foreign defense capabilities in bilateral defense relations with the United States

Table A.1 (continued)

	Output Indicators	Outcome Indicators
5. Promote stability and democracy	1. Number of meetings and activities that explicitly emphasized promotion of democracy and stability, including good governance, accountability, institution building, etc. 2. % change over previous year in number of meetings and activities that explicitly emphasized promotion of democracy and stability	1. Forum meetings and activities expanded formal and informal networks that provide mechanisms and channels to address issues in democracy and stability 2. Formal and informal networks associated with standing forums were leveraged to support democracy and stability
6. Assure allies	1. Number of forum meetings and activities that explicitly emphasized assistance to improve defense capabilities for non-forum members 2. % change over previous year in number of forum meetings and activities that explicitly emphasized assistance to improve defense capabilities for non-forum members 3. Number of of U.S.-sponsored or U.S.-organized forum meetings and activities 4. % change in number of U.S.-sponsored or U.S.-organized forum meetings and activities	1. Forum meetings and activities expanded formal and informal networks that provide mechanisms and channels to address issues in bilateral defense relations 2. Formal and informal networks associated with standing forums were leveraged to support bilateral defense relations
7. Improve non-military cooperation	1. Number of forum meetings and activities that explicitly promoted cooperation between the United States and other forum members 2. % change in number of forum meetings and activities that explicitly promoted cooperation between the United States and other forum members	1. Forum meetings and activities expanded formal and informal networks that provide mechanisms and channels to address issues in bilateral defense relations 2. Formal and informal networks associated with standing forums were leveraged to support bilateral defense relations

Table A.1 (continued)

	Output Indicators	Outcome Indicators
8. Establish relations	1. Number of forum meetings and activities that had explicit outreach to non-forum members, e.g., inviting them to sit in as observers 2. % change in number of forum meetings and activities that had explicit outreach to non-forum members 3. Forum meetings and activities provided opportunity for U.S. to expand institutional points of contact with foreign countries	1. Opened new foreign institutional contact for the Army

Bibliography

Axelrod, Robert, *The Evolution of Cooperation*, New York: Basic Books, 1984.

Bourne, M., and A. Neeley, "Cause and Effect," *Financial Management*, September 2002, pp. 30–31.

Brignall, Stan, "The Unbalanced Scorecard: A Social and Environmental Critique," available at http://www.performanceportal.org (as of November 17, 2003).

Bush, President George W., *The National Security Strategy of the United States of America*, Washington, D.C.: The White House, September 20, 2002.

Corcoran, Dr. Andrew, email discussion on British MOD efforts to measure effectiveness of military international activities, March 28, 2003.

Fearon, James D., "Signaling Foreign Policy Interests," *Journal of Conflict Resolution*, Vol. 41, No. 1, February 1997, pp. 68–90.

Fitzgerald, Lin, and Bitten Bringham, "Managing Regulatory Performance Measures," *Performance Measurement Association Newsletter*, Vol. 1, No. 4, October 2001, p. 22.

George, Alexander, and Tim McKeown, "Case Studies and Theories of Organizational Decision Making," in Robert Coulam and Richard Smith, eds., *Advances in Information Processing in Organizations*, Greenwich, Conn.: JAI Press, 1985, pp. 43–68.

Goffman, Erving, *Strategic Interaction*, Philadelphia, Penn.: University of Pennsylvania Press, 1969.

Headquarters, Department of the Army, *Army International Activities Plan Fiscal Years 2007–2008*, Washington, D.C., August 13, 2004.

Hogg, Michael A., and Dominic Abrams, *Social Identifications: A Social Psychology of Inter-group Relations and Group Processes*, London: Routledge, 1998.

Howard, Peter, "The Growing Role of States in U.S. Foreign Policy: The Case of the State Partnership Program," *International Studies Perspective*, Vol. 5, No. 2, May 2004, pp. 179–196.

Kaplan, Robert, and David Norton, "The Balanced Scorecard: Measure That Drive Performance," *Harvard Business Review*, January–February 1992, pp. 71–79.

Kaplan, Robert, and David Norton, "Putting the Balanced Scorecard to Work," *Harvard Business Review*, September–October 1994, pp. 134–147.

King, Gary, Robert Keohane, and Sidney Verba, *Designing Social Inquiry: Scientific Inference in Qualitative Research*, Princeton, N.J.: Princeton University Press, 1994.

Lee, Thomas W., *Using Qualitative Methods in Organizational Research*, Thousand Oaks, Calif.: Sage Publications, 1999.

Meyer, Marshall, *Rethinking Performance Measurement: Beyond the Balanced Scorecard*, Cambridge, U.K.: Cambridge University Press, 2003.

Moroney, Jennifer D.P., "Western Security Cooperation with Central Asia: CIS or the European Security Order?" in Graeme P. Herd and Jennifer D.P. Moroney, *Security Dynamics in the Former Soviet Bloc*, London, U.K.: Routledge/Curzon, 2003.

Moullin, M., *Delivering Excellence in Health and Social Care*, Buckingham, U.K.: Open University Press, 2002.

Neely, Andy, *Measuring Business Performance*, London, U.K.: Profile Books, 1998.

North, Douglas, *Structure and Change in Economic History*, New York: W. W. Norton, 1981.

Patton, Michael Quinn, *Utilization-Focused Evaluation*, 3rd ed., Thousand Oaks, Calif.: Sage Publications, 1997.

Patton, Michael Quinn, *Qualitative Research and Evaluation Methods*, 3rd ed., Thousand Oaks, Calif.: Sage Publications, 2002.

Performance-Based Management Special Interest Group, *The Performance-Based Handbook: Establishing an Integrated Performance Measurement System: Vol. 2,* September 2001, available at http://www.orau.gov/pbm (as of November 17, 2003).

Performance-Based Management Special Interest Group, *Serving the American Public: Best Practices in Performance Management,* June 1997, available at www.orau.gov/pbm/links/npr2.html (as of November 17, 2003).

Pollanen, Raili, and Nicola M. Young, "The Use and Effects of Performance Measures in Local Government," *Performance Measurement Association Newsletter,* Vol. 1, No. 4, October 2001, pp. 10–11.

Priest, Dana, *The Mission: Waging War and Keeping the Peace with America's Military,* New York: W. W. Norton, 2003.

Office of Management and Budget, "Program Assessment Reporting Tool," available at http://www.whitehouse.gov/omb (as of November 17, 2003).

Schelling, Thomas, *The Strategy of Conflict,* Cambridge, Mass.: Harvard University Press, 1960.

Schelling, Thomas, *Arms and Influence,* New Haven, Conn.: Yale University Press, 1972.

Sedecon Consulting, *Strategic Performance Measurement,* Espoo, Finland, 1999.

Spense, Mark, *Market Signaling,* Cambridge, Mass.: Harvard University Press, 1974.

Sugden, Robert, "Thinking as a Team: Towards an Explanation of Non-Selfish Behavior," *Social Philosophy and Policy,* Vol. 10, No. 1, Winter 1993, pp. 69–89.

Sugden, Robert, "Team Preferences," *Economics and Philosophy,* Vol. 16, No. 2, 2000, pp. 175–204.

Szayna, Thomas, et al., *Improving Army Planning for Future Multinational Coalition Operations,* Santa Monica, Calif.: RAND, MR-1291-A, 2001.

Szayna, Thomas, et al., *Improving the Planning and Management of U.S. Army Security Cooperation,* Santa Monica, Calif.: RAND, MG-165-A, 2004.

Tajfel, Henri, *Human Groups and Social Categories*, Cambridge, U.K.: Cambridge University Press, 1981.

Taw, Jennifer Morrison, *The Effectiveness of Training International Military Students in Internal Defense and Development: Executive Summary*, Santa Monica, Calif.: RAND, MR-172-USDP, 1993.

United Kingdom National Audit Office, *Measuring the Performance of Government Departments*, HC 301 Session 2000–2001, London, March 22, 2001.

U.S. Department of the Army, "Strategic Readiness Reporting," available at http://www.army.mil/aps/2003/realizing/readiness/reporting.html (as of November 17, 2003).

U.S. Department of Defense, *Quadrennial Defense Review Report*, Washington, D.C.: September 30, 2001, available at www.defenselink.mil/pubs/qdr2001.pdf (as of November 14, 2003).

U.S. Department of Defense, *Department of Defense Security Cooperation Guidelines* (U), Washington, D.C., April 30, 2003 [not available to the general public].

U.S. Department of Energy, *Guidelines for Performance Measurement*, Washington, D.C., 1996.

U.S. Department of State, "FY 2004 Performance Plan," 2003, available at http:// www.state.gov/m/rm/rls/perfplan/2004/ (as of November 17, 2003).

U.S. General Accounting Office, *Effectively Implementing the Government Performance Results Act: Executive Guide*, GA/GGD-96-118, Washington, D.C., June 1996.

U.S. General Accounting Office, *Managing for Results: Analytic Challenges in Measuring Performance*, GAO/HEHS/GGO-97-138, Washington, D.C., May 1997.

Van Evera, Stephen, *Guide to Methods for Students of Political Science*, Ithaca, N.Y.: Cornell University Press, 1997.

Wendt, Alexander, *Social Theory of International Politics*, Cambridge, U.K.: University of Cambridge Press, 1999.

Wholey, Joseph S., et al., eds., *Handbook of Practical Program Evaluation*, San Francisco, Calif.: Jossey-Bass Inc., 1999.